Principles
and Techniques
of Scanning Electron
Microscopy

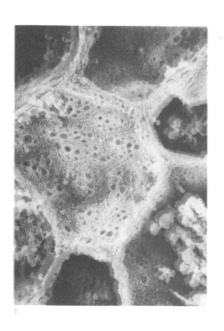

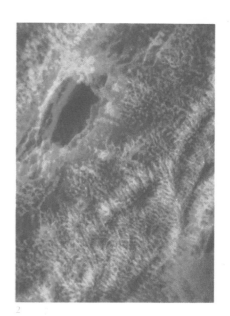

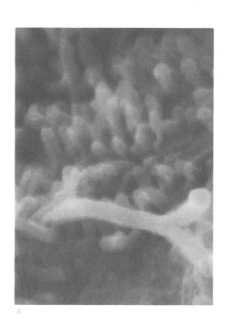

Principles and Techniques of Scanning Electron Microscopy

Biological Applications
Volume 1

Edited by

M.A. Hayat
Department of Biology
Kean College of New Jersey
Union, New Jersey

 VAN NOSTRAND REINHOLD COMPANY
NEW YORK / CINCINNATI / TORONTO / LONDON / MELBOURNE

Van Nostrand Reinhold Company Regional Offices:
New York   Cincinnati   Chicago   Millbrae   Dallas

Van Nostrand Reinhold Company International Offices:
London   Toronto   Melbourne

Published by Van Nostrand Reinhold Company
450 West 33rd Street, New York, N.Y.   10001

Published simultaneously in Canada by Van Nostrand Reinhold Ltd.

15 14 13 12 11 10 9 8 7 6 5 4 3 2 1

Library of Congress Cataloging in Publication Data

Hayat, M A
    Principles and techniques of scanning electron microscopy.

    Includes bibliographies.
    1. Scanning electron microscope—Technique.
I. Title.   [DNLM: 1. Microscopy, Electron, Scanning.
QH212.E4 H413p]
QH212.S3H38   502'.8   73-16011
ISBN 0-442-25677-9 (v.1)

PREFACE

The scanning electron microscope has some significant advantages over both the light microscope and the transmission electron microscope. Scanning electron microscopy provides a greater depth of field than that provided by light microscopy at the same magnification. The usefulness of the former for the examination of biological specimens having irregular surfaces is therefore apparent. Also, scanning electron microscopy provides three-dimensional images, whereas transmission electron microscopy provides essentially two-dimensional images because it must utilize flat, thin sections. Another advantage of the scanning electron microscope is that it provides a clear view of much larger specimens than does the transmission electron microscope.

The rapid progress in both the methodology of specimen preparation and instrument design during the past few years is responsible for the wider application of scanning electron microscopy to biomedical areas. The introduction of critical point drying to the methodology is one striking example of the relatively recent improvements of major importance. The development of the high-resolution scanning electron microscope is probably the single outstanding advance. The extent and significance of its potential use in the examination of biological specimens are extraordinary.

The scanning electron microscope has already been employed for studying the distribution of enzymes and antigens on the surfaces of cells and tissues, the elemental composition of the specimen, whole or fractured, quick-frozen specimens in the cold stage, and living organisms. The science of scanning electron microscopy is advancing rapidly toward the goal of total characterization of the specimen under study.

Since a book containing detailed methods for the preparation of a wide variety of biological specimens for scanning electron microscopy is not available, the need for compiling these methods is apparent. To that end, a three-volume series has been planned. Volume 1 presents methods applicable to botanical specimens, except for the first six chapters which are of a general nature. Subsequent volumes will concentrate on the processing of animal specimens.

Because it is difficult for a single author to satisfy the needs of all readers in a rapidly developing field of research, the collective experience of as large a group of scientists as possible was solicited. These scientists represent six countries, and have varied backgrounds in the biomedical and biophysical areas. As a result of this approach, a most extensive compilation of methods, developed and used by a large group of competent scientists, has been achieved. The series represents the experience and judgment of the scientists who have been using these methods in the course of their professional work.

The methods presented have been tested for their reliability, and are the best of those currently available. The instructions for the preparation and use of various solutions, media, coats, and apparatus are straightforward and complete, and should enable the worker to prepare his own specimens without outside help. It is suggested that the entire procedure be read and necessary solutions and other media prepared prior to undertaking the processing. Each chapter is provided with an exhaustive list of references with complete titles; and full author and subject indexes are included at the end of the volume.

The reader is told what is known as well as the gaps in our knowledge. Alternative procedures, points of disagreement, and potential research areas are pointed out. The focus is on those aspects of methodology which have contributed most to the satisfactory processing of the specimens in the past and are likely to be in the vanguard of improvements to come. The book is intended not only for teachers and scientists but also for students, technicians, and research workers not familiar with the techniques.

The methods presented are, however, subject to modifications depending upon the objective of the study and the worker. It is hoped that this volume will contribute significantly to the development and standardization of methodology. Considering the speed of progress in methodology during the past few years, rapid improvements in the presently available methods are to be expected.

I am grateful to Michael Nemanic for his continued help during the preparation of this volume. I acknowledge with pleasure the cooperation and friendship offered by George Narita and Alberta Gordon of Van Nostrand Reinhold Company.

M. A. HAYAT

CONTENTS

3 CRYOTECHNIQUES
Tokio Nei

Contributors to This Volume

E. A. Baker

Lee A. Bulla, Jr.

J. Temple Black

Arthur L. Cohen

Karl Borgin

Joseph F. Gennaro, Jr.

L. G. Briarty

P. J. Holloway

H. F. Howden

Michael Nemanic

Cletus Kurtzman

Ann W. Nickerson

L. E. C. Ling

Barbara J. Panessa

Tokio Nei

Keiichi Tanaka

Contents of Volumes 2-4

1. THE SCANNING ELECTRON MICROSCOPE: OPERATING PRINCIPLES

J. Temple Black

Industrial Engineering Department, University of Rhode Island,
Kingston, Rhode Island.

INTRODUCTION

The rapid advances in the development of the scanning electron microscope (SEM) during the past decade have made it an important tool for the examination of surfaces. A dramatic example of what this instrument can accomplish is illustrated in Fig. 1–1, which shows the surface details of part of a dime. Because of the depth of field or three-dimensional quality of the image obtained, electron micrographs produced by this instrument can be easily interpreted in terms of what the microspace really looks like. Micrographs produced by optical microscopes or transmission electron microscopes (TEM) are primarily two-dimensional images, the former because of its narrow depth of field, and the latter because the specimens must be flat, thin slices.

The SEM produces micrographs by scanning the surface of a specimen with a small electron probe (a beam of electrons) synchronously with an electron beam in a cathode ray tube. The contrast is due to topographical variations and atomic number differences in the specimen. The signals produced by the probe-specimen interaction are used to modify the intensity of the beam in the cathode ray tube.

Because the interactions between the electron probe and the specimen are diverse, the instrument has considerable flexibility. Its potential application can be extended to almost every field of scientific endeavor

1

Fig. 1–1. A scanning electron micrograph of the surface of a dime. The letter "s" in the phrase "In God We Trust" is imaged by an interaction of an electron probe with the surface of the dime. ×90.

concerning surfaces, and is limited primarily by specimen preparatory procedures. However, one must remember that these remarkable pictures were produced by an electron beam interacting with the specimen surface. This chapter will explain the operation of the microscope and demonstrate its application to the study of a wide variety of biological specimens; the nature of electrons and their interaction with matter are also considered.

ELECTRONS AS IMAGE PRODUCERS

Electrons are small, charged particles, which also exhibit a wave nature in the free state. Their wavelength, which depends upon the inverse of their momentum, is very short (less than 1 Å) at the voltages used in electron microscopy. Because electrons are charged particles, they can be deflected by electrostatic or electromagnetic fields. Thus they can be accelerated, bent in their flight path, and focused so that they can be used in electron optics as photons are used in light optics. In general, however, electron optical systems have much greater resolution and depth of field than those possessed by light optical systems. Electron microscopes can be designed to image internal structures as well as external surface morphology. Because we deal with a world of matter and energy conversion, instruments capable of resolving matter at the atomic level are imperative to the understanding and the advancement of science, technology, and medicine.

Knoll and Ruska (1932) are credited with the development of the first magnetic lens electron microscope which is generally recognized to be the prototype of the existing conventional TEM. This development was preceded by the following three fundamental discoveries concerning electrons:

(1) The existence of electrons was proved (Thompson, 1897).

(2) The hypothesis was advanced that electrons have a wave nature while still being charged particles (de Broglie, 1923).

(3) Electrical and magnetic fields can operate as lenses on charged particles by generating a symmetrical lens action (Busch, 1926).

The history and development of TEM can be found in standard books by Heidenreich (1964) and Grivet (1965). Almost at the same time that the TEM was being developed, Knoll (1935) suggested that a different kind of electron microscope could be developed by focusing a fine, scanning beam of electrons on a surface and recording the emitted current as a function of the position of the beam. Knoll, along with von Ardenne (1938), is credited with building the first SEM. Von Ardenne described an SEM which used two magnetic lenses to demagnify the beam, employing thin films, rather than solid surfaces, for specimens, and using the transmission properties of the beam for structure determinations. This microscope was thus the forerunner of the scanning transmission electron microscope (STEM), which is currently being developed by Crewe (1971) and others.

Von Ardenne suggested that secondary electrons be collected from the tops of opaque surfaces, amplified, and used to modulate the grid of a cathode ray tube. Zworykin et al. (1942) developed an improved version of the SEM. But these instruments did not show the depth of field and resolution associated with presently available instruments because the signal collection and amplification processes were poorly understood and developed. The SEM developed by Zworykin et al. (1942) used electrostatic lenses (two) to produce a 500 Å probe, but oil contamination led to the termination of this project.

Brachet (1946) discussed the theoretical limit of resolution achieved by the SEM and agreed with the theory presented by Zworykin et al. (1945) that the attainment of 100 Å resolution might be possible. Leauté and associates in 1946 built an SEM capable of resolving up to a few microns. Davoine (1957) reported the construction of an SEM capable of resolving ∼2 μm for the study of secondary emission from stressed metals. Davoine *et al.* (1960) used this instrument for cathodoluminescence displays.

Meanwhile, Oatley started the work in 1948 that eventually led to the construction of the first commercial SEM in 1965. McMullen (1953) solved the noise problem in the electron collection system by utilizing a secondary electron multiplier which could operate in a demountable system. The multiplier used was the beryllium-copper dynode system developed by Baxter (1949). This instrument employed electrostatic lenses, double deflection of the beam, long scanning durations, and long-persistence CRTs for direct display. Smith (1956) added gamma controls, stigmation correction, micromanipulation of the specimen in the microscope, and a water vapor cell. He extended the applications (a hot stage and cathodoluminescence from phosphor) of SEM while achieving a resolution of ∼250 Å.

The above-mentioned early developmental work was followed by additional improvements in instrumentation and advances in the understanding of the factors affecting contrast and resolution. The scintillator-photomultiplier system of electron detection commonly employed today was developed by Everhart and Thornley (1960). They found that as long as the accelerating voltage applied to the detector was greater than 10 kV, a substantially noise-free amplification of the initial signal could be obtained. Pease (1963) demonstrated a beam diameter of 50 Å and micrograph resolution of ∼100 Å, which are considered to be satisfactory operating conditions even today. The history of the development of the first commercial SEM has been reviewed by Oatley *et al.* (1965).

BASIC ELECTRON OPTICAL SYSTEMS

At present, there are three basic electron microscopy systems commercially available.

(1) The TEM or CTEM is the oldest and most highly developed system. This instrument images internal structure by transmitting electrons through thin sections of the specimen and recording the image on a film. It is capable of practical resolution in the order of 2 to 3 Å, and operates at accelerating voltages of 40 keV to 1 or 3 meV.

(2) The SEM employs the interaction products images of the topography and composition of the surface. It acquired its name because a pencil of electrons is "scanned" over the surface in a raster pattern and images are constructed on a point-by-point basis on the face of a cathode ray tube.

(3) The STEM, the newest instrument, possesses the combined char-

acteristics of the TEM and SEM. A pencil beam of electrons is scanned over a thin film of the specimen, and electrons are transmitted through the thin film. The transmitted electrons are collected and imaged on a point-to-point basis as in the SEM, while in the TEM the specimen is illuminated everywhere simultaneously.

THE TRANSMISSION ELECTRON MICROSCOPE

In this instrument, the thickness of the specimen penetrated by electrons is much thinner ($\sim$0.1 μm) than those examined with an optical microscope. The sample is continuously illuminated over the entire field of view by the collimated beam of electrons coming from an electron gun. The electron source has traditionally been a heated tungsten wire. The electrons, after emission, are accelerated by a voltage drop of 40,000 to 125,000 volts, and are shot down the evacuated microscope column. The thin section in the microscope serves to scatter this nearly monochromatic beam of electrons.

The primary interactions that occur between the electrons and the atoms of the specimen are:

(1) Elastic scattering—primary electrons are deflected by atoms but do not lose energy.

(2) Inelastic scattering—primary electrons deflect and lose energy.

(3) No scattering—primary electrons pass through the specimen undeflected.

Other lenses (objective, intermediate, and projector) below the sample then focus the elastic and unscattered electrons (which will have the same energy and wavelength) on a screen or sheet of film while magnifying the image. The inelastically scattered electrons tend to degrade the quality of the image (as chromatic aberration) because they are out of phase with the image-forming electrons. Their scattering angles are, on the average, smaller than the angles made by elastically scattered electrons, so that they are not completely filtered out by small objective apertures and reach the final image position (the screen or film) as a background noise. The inelastically scattered electrons are out of phase with the rest of the image-forming electrons because they have lost energy and have suffered a concomitant change in wavelength. The transmission image is largely a measure of the scattering potential of the atomic arrangements in a thin specimen.

The ultimate resolution capability (r) is arrived at through a compromise between the spherical aberration inherent in electromagnetic lenses and the diffraction condition due to the wave nature of the electron expressed by the Abbe equation (Eq. 1–1) (Jenkins and White, 1951).

$$r = \frac{.61\lambda}{\alpha} \qquad (1\text{-}1)$$

In Eq. 1, α = alpha = half the aperture angle of the illumination, and λ = lambda = the wavelength of the electron beam, while r = the radius of a disk of illumination at the image plane (called the Airy Disk). Thus according to Eq. 1-1, even for a perfect lens, a point source of illumination will be imaged as a disk of radius r. The resolution capability is further degraded by lens aberrations.

One can see, however, that improving the resolution requires the reduction of r. Two options are immediately obvious: reduce the wavelength of the illumination by increasing the accelerating voltage, or increase the aperture angle α by using apertures of larger diameter or by reducing the distance between the sample and the aperture. However, to increase α by increasing the aperture size is to incur rapid increases in spherical aberrations (bad lens focusing) and other lens errors, so that resolution capability is lost faster than it is gained. The development of high-voltage TEMs, where the electrons are accelerated through voltage drops of 500,000 to 1,000,000 volts or more, somewhat improves resolution because the electron wavelength is reduced (Eq. 1-2).

$$\lambda = \frac{12.3}{eV} \qquad (1\text{-}2)$$

More important, these high-energy electrons can penetrate thicker specimens, enabling HVTEMs to resolve internal structure in thick specimens (1 to 5 μm) with good resolution. Unfortunately, such systems are very expensive to purchase and maintain, but they are nonetheless increasing in number throughout the world. The ability to look through thick sections greatly reduces the specimen preparation requirements for conventional transmission electron microscopy while eliminating many image interpretation problems encountered with ultrathin sections.

THE SCANNING ELECTRON MICROSCOPE

The SEM (Thornton, 1968; Everhart and Hayes, 1972) can reveal topographical details of a surface with clarity and detail which cannot be obtained by any other means. This instrument can also detect surface potential distributions, subsurface conductivity, surface luminescence, surface composition, and crystallography. It can resolve topographical details of less than 50 Å with a depth of focus 500 times that of an optical microscope at equivalent magnifications. The SEM operates typically at resolutions in the order of 100 Å.

Operational Principles

The SEM image can be produced by any signal generated by the interaction of a finely focused primary beam of electrons as it is scanned over the specimen (Fig. 1–2). As indicated in this figure, a narrow beam of electrons is produced by successive electromagnetic condensor lenses which place a small spot of electrons on the specimen ($\sim$500 to 50 Å in diameter). The primary beam (also called the electron probe) penetrates the surface and produces a variety of signals, any of which can be used to generate micrographs.

Secondary electrons (e_2^-) have low energy, and can escape only from a very thin layer of the surface (50 Å). Many of them are reabsorbed by the specimen itself. The main effective area of secondary electron emission is very close to the specimen-incident beam interface, so that the probe size is the dominant feature in the resolution capability of a SEM. The specimen-beam interaction also produces backscattered electrons (these are elastically scattered electrons), specimen currents, photons, Auger electrons, and X rays which are characteristic of the probed specimen. In addition, induced currents can be developed in semiconductive specimens. These signals are then used to produce images (micrographs) of the surface in the following way.

The basic machine is composed of an electron gun which provides a beam of electrons with energies of from 1 to 50 keV. The electrons are accelerated past two or more condensing lenses that demagnify the beam into a small-diameter probe, which is then scanned over the specimen. In order to scan the specimen with the probe, deflection coils are placed between the last two lenses (or within the final lens) to deflect the beam in a rectangular pattern over the sample. The scan generator, which produces sweep signals to the column deflection coils, at the same time operates deflection coils in the SEM's cathode ray tubes. Because of this synchronism, there is a one-to-one correspondence between the position of the electron beam on the specimen and that of the spot on the CRT.

The intensity of the CRT spot can be controlled by the strength of a signal reaching the control grids (G) in the CRTs. The grid in a CRT is really a metal cap which lies between the electron source and the phosphor face of the CRT. The strength of the electric field placed on this grid determines the volume of electrons which can pass through it to the CRT face. The electron beam in the CRT is modulated in intensity by the strength of signal sent to the CRT grids from the detector. Therefore the brightness of the spot of the face of the tube is controlled by the amount of electrons liberated from the specimen surface by the inter-

IMAGE FORMING IN THE SEM

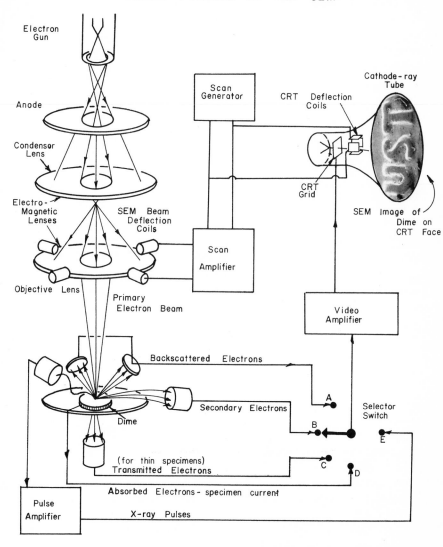

Fig. 1–2. Schematic diagram of the SEM. Various signals produced by the beam-specimen interaction can be used to generate micrographs which are obtained by photographing the face of a cathode ray tube (CRT). See Fig. 1–1.

action of the electron probe with the specimen. Because a major part of the electrons reaching the detector are secondary electrons, and because they are easily captured by depressions in specimen topography, a specimen hole will appear black and a specimen hill will appear white on the face of the SEM CRT. Thus the SEM image is built up on the face of the CRT on a point-to-point basis, and a photograph of the image on the face of the tube is a scanning electron micrograph.

Surface Emission Considerations

The demagnification of the beam by two or three magnetic lenses is a necessary operation, since for heated tungsten filaments, the electron beam at its source may be 20 to 50 μ in diameter. Obviously a large percentage of the source emission is lost or screened out by the apertures and lenses; thus the beam current (I_p) in the SEM is reduced from about 10^{-4} amp at the gun to 10^{-10} or 10^{-12} amp at the specimen (for heated tungsten filament sources). The number of electrons generated at the specimen surface is related to the beam current. The electrons intercepted by the demagnification lenses result in this beam current reduction. A beam of 10^{-4} amp provides approximately 10^{15} electrons per second, while a beam of 10^{-10} amp provides about 6×10^6 electrons per second. (The fewer the number of primary electrons into the specimen, the less the volume of emission (per unit time) coming from the sample.)

The amount of current that can be focused into the final spot determines the strength of the emission signal from the specimen surface; the signal must be large enough so that after amplification it has sufficient strength to overcome the electronic noise inherent in the electron collection and amplification system. Too much demagnification of the spot on the specimen (below about 50 Å) reduces the number of electrons available to interact with specimen, and consequently reduces the number of electrons emitted by the surface by the probe point. The amount of current that can be supplied to the final spot is directly proportional to the current density of the source, and so special high-density electron sources have been developed which greatly improve SEM performance (see "Electron Gun Sources," below) in that they can put more electrons into a given spot diameter. In the typical SEM system employing heated tungsten filaments, the probe current can be increased by inserting a larger aperture in the beam, thus increasing the convergence angle of the probe. While this improves the signal-to-noise (S/N) ratio, it also increases the spot size (which decreases resolution) and decreases the depth of field. Thus the aperture angle of the beam is usually kept to half a degree or less in the final lens. Typically, SEMs use 50 μm, 100 μm, or 200 μm final apertures with an aperture to specimen distance of 15 mm.

Electron Collection and Imagery

The electron emission from a surface depends upon the following factors:

(1) The accelerating voltage (typically 1 to 40 kV) or the incident energy of the probe electrons.

(2) The surface morphology and the angle which the primary beam makes with the surface.

(3) The atomic density (Z) of the site, since Z influences the distribution of the incident beam in the specimen—that is, the penetration and absorption of the primary beam and scattering of the electrons in the solid are greatly influenced by the atomic weight of the specimen.

(4) The surface chemistry and crystallography, which influence the potential barrier of the surface.

(5) Local charge accumulations on the surface.

The relative intensity of the emission from a surface has a typical energy profile, as shown in Fig. 1–3. Changes in the accelerating potential alter the depth of penetration of the probe into the specimen surface, which, in turn, influences the intensity distribution. However, it seems that within the 10 to 40 kV range, the variation of the angle between

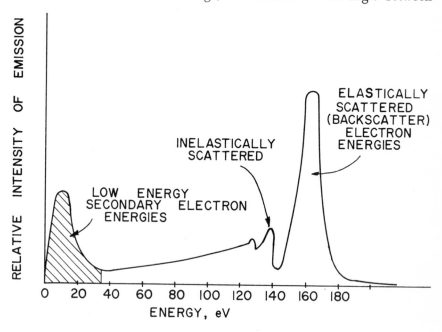

Fig. 1–3. Typical energy distribution of emitted electrons in terms of relative yield as a function of their energy upon emission. The secondaries have an average value of ∿5 eV.

the incident or primary beam and the surface of the specimen plays the major role in emission signal strength variation. *Thus the topography of the specimen surface determines the image formation and contrast whether secondary electrons or reflected electrons are being used.*

In Fig. 1–4, for example, different planes have different contrast. An alteration of only 1 or 2 degrees can cause an appreciable change in brightness of the final image. In rougher specimens, additional contrast can result from changes in orientation of the specimen with respect to the collector. That is, contrast may result from one area shading another area or the specimen being tilted so that an electron in a pit or hole cannot escape to the collector. This is called "specimen contrast," and can be observed in Fig. 1–4, where crystal *C* is shaded from the collector by crystal *B*.

As shown in Fig. 1–5, the interaction of the beam with the surface

Fig. 1–4. A scanning electron micrograph of silver chloride crystals. The top surface (*T*) is making an angle of ∽45° with the primary electron beam. Other surfaces lying at different angles show different levels of contrast. The edges (*E*) of the crystal are brighter, owing to excessive secondary emissions.

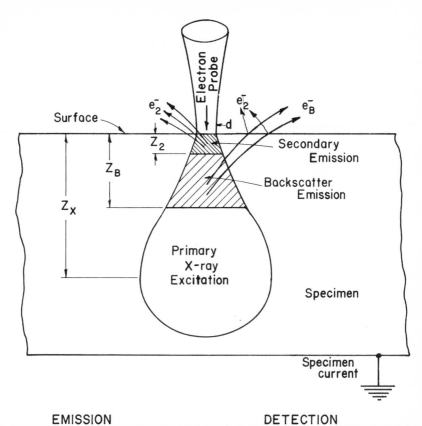

EMISSION		DETECTION
Secondary	Electrons	Scintillator, Photomultiplier,
Backscatter	Electrons	and Amplifier
Characteristic	X-ray	Spectrometers – Li Drifted Si Semiconductors
Cathodoluminescence		Monochromator - Photomultiplier

Fig. 1–5. Typical electron penetration volume in a material with a low Z number probed with a beam 25 to 45 kV. As kV is reduced or Z increased, the teardrop changes into a hemisphere. (See Kimoto 1972)

produces signals from different levels below the surface. Typically, $Z_2 = 50$ to 500 Å, $Z_B = 1000$ Å to 1 μm, and $Z_x = 5000$ Å to 5 μ for a spot of diameter d. As stated earlier, this distribution of signal intensity is a function of the composition of the specimen material and the energy of the incident probe. The larger the energy of the probe V and the smaller the specimen's atomic number Z, the deeper will be the penetration d_p. That is,

$$d_p \propto \frac{W_a V_0^2}{Z \rho} \qquad (1\text{-}3)$$

where W_a = atomic weight, ρ = material density. Diffusion of the electrons in the transverse directions increases compared to d_p as Z increases or V decreases, so that the teardrop changes to a hemisphere.

Sharp edges or corners will appear brighter or lighter than their adjoining surfaces because of the additional secondary electron emission caused by the glancing incidence of the electron probe (Fig. 1–4). Thus the interaction of the beam with the surface is influenced by the atomic weight of the atoms in the surface, the topography of the surface, and the energy of the probe itself. Heavy metals coupled with accelerating voltages of 10 to 30 KeV produces excellent imaging in the SEM; this is one reason for coating biological material with gold or platinum instead of aluminum or carbon.

Electron Collection. The electron collector for secondary electrons is positively (40 to 200 V) charged, and attracts the relatively low-energy secondary electrons (e_2^-) emitted by the specimen (Fig. 1–6). The backscattered electrons, which have undergone relatively little loss in energy, are not appreciably deflected by a potential difference of a few hundred volts and therefore do not enter the scintillator collector unless the latter happens to lie directly in their path. The backscattered electrons can generate secondary electrons at the point where they emerge from the surface and at surfaces other than the specimen. These secondaries will contribute to the final signal. For example, indirect secondaries liberated by reflected electrons striking the bottom of the final lens or the inside of the collector can make up 30% of the output signal.

The ratio of secondary to backscattered electrons in the signal can be varied by geometrical arrangements of the specimen and the collector and by the collector potential itself. That is, the collector voltage or the accelerating voltage in the cup may be turned off so that images are formed only by the backscattered electron emission, but the resolution in such arrangements is poorer than normally obtained with both secondary and backscattered emission.

The backscattered electrons are better detected by a pair of detectors placed on the bottom of the final lens, as shown in Fig. 1–6. Image contrast for the backscattered electrons is determined essentially by the intensity of these elastically scattered electrons; intensity, in turn, de-

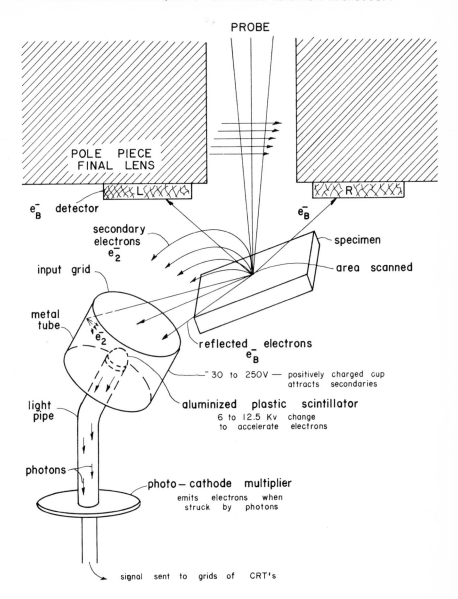

Fig. 1–6. Collection and detection of secondary and backscattered or reflected electrons from the specimen-probe interaction region. Secondaries and some backscattered electrons are collected by the scintillator arrangement, while the pair detector picks up the backscattered or high-energy types. In future microscopes, these signals will probably be integrated or mixed electronically on a point-by-point basis to build a single image.

pends upon the average atomic number of the specimen and the incident angle of the primary beam on the specimen. Thus, if the signal from the left detector is added to that of the right detector, topography differences are canceled out and composition differences are observed. If the difference between the two signals is used, topography will be emphasized. An example of this effect is shown in Fig. 1–7. The backscatter detectors need not be positioned as shown in Fig. 1–7, but can be mounted elsewhere in the specimen chamber for special applications.

The secondary electrons collected by the detector +200 to 400 V field are then accelerated into an aluminum-coated scintillator by an additional 10 to 12.5 kV field. The electrons striking the scintillator produce photons which travel down a light pipe; the aluminum coating then serves as a mirror or reflector surface to direct the photons toward a photomultiplier, which produces a photocurrent that is amplified and used to intensity modulate the CRT brightness.

Deterioration of the aluminum scintillator coating is a prime cause of the loss of signal in the SEM. Recoating by vacuum evaporation requires a very clean, oil-vapor-free vacuum, or the scintillator performance will fall off rapidly (2 weeks to a month) compared to normal performance life of 3 months. High vacuum levels (10^{-9}) with LN_2 trapping are suggested for scintillator coating. Aluminum is evaporated while rotating the pipe during this process; the thickness obtained is critical. It is recommended that this part of the instrument be maintained in the same manner as filaments (i.e., purchased from SEM company and kept in stock and changed as required by S/N levels). Generally, any signal collected from the specimen can be used to intensity modulate the CRT beam. Most SEM systems allow for multiple viewing of different images or easy switching from one signal to another.

In summary, the current reaching the collector varies as the primary electron beam scans the surface of a specimen in synchronization with the beam in the CRT; the strength of this emission signal determines the brightness of the spot on the face of the CRT. Although one cannot assume that this image will be identical in appearance to that obtained with an optical microscope, it is similar.

On a flat surface, tilting the specimen with respect to the primary beam causes the spot to elongate in the direction of the tilt. The signal-to-noise ratio and the resolution are usually improved by tilting, but the depth of field requirements are increased. Inserting smaller apertures (50 to 100 μm) into the primary beam below the final lens causes the beam to become more collimated and thereby to have a greater depth of field at the specimen level; there is also some loss in signal strength caused by decreased beam currents. The less the specimen is tilted, the

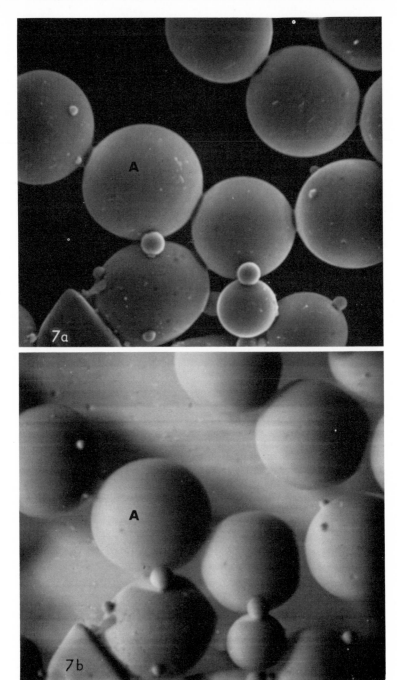

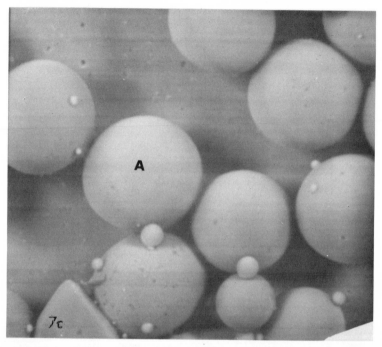

Fig. 1–7. Glass spheres imaged by normal secondary mode in 7a and by e detectors in 7b (topography mode) and 7c (composition mode). Because there is no change in composition in the specimen, the spheres appear flat and lack contrast in (c). Notice that the bridges between the spheres are easily seen in (a), but not in (b) and (c). The micrographs were obtained with the surface of the spheres being perpendicular to the probe. Sphere A is ∽100 μm in diameter. Spheres are gold coated.

flatter the field of examination will be. Thus larger apertures (200 μm) can be used, resulting in a stronger signal.

Fig. 1–8 demonstrates these effects by using a TEM copper specimen grid as a subject. In Fig. 1–8a, the grid is examined while lying perpendicular to the beam. Tilting the specimen to 45° (Fig. 1–8b) foreshortens the image and leaves the upper and lower portions out of focus when a 200 μm final aperture is used. A smaller aperture (100 μm) increases the depth of focus by collimating the beam and brings the entire specimen into focus.

The working distance in the SEM is defined as the distance from the bottom of the final lens to the midplane of the field scanned on the specimen. The working distance can be varied in most SEMs by translating the specimen in the z axis (or beam axis) of the scope. Increased working distance decreases the beam convergence angle—in other words, the beam becomes more collimated and will have a greater depth of

SECONDARY ELECTRON IMAGES OF TEM
COPPER SPECIMEN GRID AT 5 mm W.D.
Photographed at 100 sec scan rate

	MAG	TILT	APERTURE
Fig a	135	0°	200μ
Fig b	135	45°	200μ
Fig c	135	45°	100μ

Smaller aperture increases depth
of focus. Tilting foreshortens
image.

Fig. 1–8. Secondary electron images of a TEM copper specimen grid. Micrographs produced at a 5 mm working distance and photographed at a 100 second scan rate. The smaller aperture increases the depth of focus and tilting forshortens the image.

focus. Decreasing the aperture size will have a similar effect, but will reduce the signal strength. The resolution will become limited by diffraction effects and the signal-to-noise ratio for very small convergence angles. An increase in the z distance also decreases resolution because the final lens is less excited.

Magnification

Magnification in the SEM is the ratio of an adjustable current applied to the deflection coils in the scope to the current applied to the deflection coils in the CRT. The deflection coils control the size of the area scanned on the surface of the specimen with respect to the size of the CRT face, which is constant in size. Magnification can be varied over a wide range while preserving focus and specimen location, since a change in magnification merely alters the area scanned on the specimen. Thus, as smaller areas are scanned, the magnification increases, and therefore a smaller diameter probe should be employed to keep the resolution level concomitant with the magnification. Because a smaller probe usually results in lower electron emission per unit time, slower scan rates (longer exposure times) during photography of images should be employed to allow the primary beam to dwell longer at a spot. This will improve the signal-to-noise ratio.

Calibration of magnification should be checked whenever one uses altering imaging systems (such as TV recording), since a different set of deflection coils in the magnification ratio will alter the real magnification of the specimen. Altering the accelerating voltage may also require recalibration of magnification. While no real standard exists for this purpose, standard TEM specimen grids (1000 mesh) mounted flat on SEM stubs yield acceptable magnification standards. Silver conducting paint is a satisfactory mounting medium for many types of specimens. Details in the low-magnification micrographs (where the grid bar spacings can be used to check dimensional accuracy) can be used at the higher magnifications to check the calibration.

Since the image is foreshortened by the tilting operation, magnification calibration should be carried out with the specimen flat or perpendicular to the beam. The square grid will also show any image distortion problems which may exist at lower magnifications. One can also gold-coat a TEM diffraction grating for magnification calibration at higher magnifications.

Display Units

Scanning electron microscopes are generally equipped with at least two CRTs for visual inspection and examination of the specimen and for photographic purposes and a regular TV monitor. The visual display CRT has a high-persistence screen for the purpose of focusing the instrument and selecting the areas of interest for photographing. When this CRT is operated at one frame per second repetition frequency, a reasonable compromise between speed of operation and noise integration is obtained.

Although the visual display tube can be photographed, this is not the usual procedure, for two reasons: (1) halation impairs definition in the final image (afterglow around the spot on the CRT screen), and is difficult to avoid; and (2) the image, as recorded on the film, will be blurred if any electrical or mechanical drifts occur during exposure extending over many cyclic frames. While some SEMs do operate in this manner, it may be better to provide for a separate photographic recording display unit wherein the specimen and the CRT screen are scanned once, occupying the entire exposure time.

The photographic screen has clear definition, short afterglow, and very little halation. Any drift will cause distortion in the image but no blurring, since the final photographic image is constructed one line at a time. For plastic deformation experiments and for nondestructive testing inspection purposes, it is advantageous to equip the SEM with a TV scanning device. The scanning rates in the SEM are matched to those of the standard TV system. Thus a conventional video tape recorder can be used for recording and reproducing the image by taping the signal delivered to the TV monitor.

There is some loss in resolution in the TV system in SEMs where tungsten filaments are used, and so lower magnifications are employed. Satisfactory picture definition is lost at a magnification of approximately 5,000 times. By using TV, focusing and specimen shifting are easier than in conventional SEM, and this accessory is essential in some SEM systems for locating specific areas of the specimen. The TV scanning device can be employed with deformation and heating stages as well as with conventional stages so that processes can be dynamically observed. The SEM does not generate a tracking pulse for the VTR, and so later playback of the tape may be troublesome.

Electron Gun Sources

The surface potential of metal acts as a step barrier to the electrons leaving a metal surface. There are four ways to overcome this barrier:

(1) Photo emission, where the surface is bombarded with photons of energy hν. This is the reverse of cathodoluminescence (see De Mets in this volume).

(2) Thermonic emission, where the metal is heated to excite the electrons.

(3) Secondary emission, where the surface is bombarded with electrons (used in image formation).

(4) Field emission, where a strong positive field pulls the electrons from the metal.

Thermonic emission and field emission are used in electron microscope guns. The tungsten (W) and lanthanum hexaboride (LaB$_6$) sources are basically thermonic types, where heating the metal increases the ion-core vibrations which induce an increase in electron collisions. This gives the electrons added kinetic energy. The addition of an electrostatic field helps suppress the height of the effective barrier. The emission J_c in amp/cm^2 at the gun as a function of absolute temperature (T) is given by

$$J_c = AT^2 e^{-b/T} \qquad (1\text{--}4)$$

where A and b are constants. J_c is called intensity or beam current density. Brightness is the flux per unit area per unit solid angle, and is given by the symbol β in amp/cm^2/steradian.

In Equation 4, b is proportional to the work function, T is the absolute temperature in degrees Kelvin, and A is an emperical constant dependent upon the material of the filament. The final beam current i is

$$i = \frac{\pi}{6} d_0^2 J_c \frac{eV}{kT} \alpha^2 \qquad (1\text{--}5)$$

where d_0 is the final beam diameter in absence of all other effects, including diffraction and aberrations. It may be given simply as

$$d_0 = \frac{D_0}{M} \qquad (1\text{--}6)$$

where D_0 is the diameter at the source and M is the magnification or by rewriting Eq. 1–5. The source diameter is the apparent source at crossover below the cathode, and is typically 50 to 100 μm, while d_0 is of the order of 50 to 100 Å.

Thermonic emission of electrons has for years been the traditional source of electrons for all types of electron microscopes. A tungsten wire, bent into a V or U shape is resistance-heated by an electric current, which is called the filament current. At a temperature of about 2900°K, the electrons in the tungsten have sufficient energy to be boiled off the filament and drawn away by an electric field. The critical parameter for an electron gun is beam brightness. It determines how much current can be focused on a given spot for a given electron optical system.

The tungsten filaments are limited in sharpness at their points by curvature of the bend in the tungsten wire. The sharper the point of a filament is, the stronger will be the field at the tip. The tip field strength determines how large an electron current can be drawn from the cathode before the mutual repulsion of the electrons in the beam begins to widen the beam significantly. Making the tungsten wire sharply pointed thus improves the brightness, but the metal in the filament evaporates at a rapid rate at the temperatures required for emission of sufficient numbers of electrons; so the life of pointed tungsten filaments is rather short. Even the tungsten hairpin types last only 30 to 60 hr at normal operating vacuums ($\sim 10^{-5}$ torr) and filament currents.

The lanthanum hexaboride (LaB_6) gun developed by Broers (1970) at IBM and the field emission gun (FE) employed by Crewe *et al.* (1968) are two alternative systems which produce marked increases in the beam current for a given spot size. Lanthanum hexaboride emits copious amounts of electrons at about 2000°K with very little metal evaporation. The rod can be ground to extremely sharp points (1 μm diameter) to combat the space charge limitation problem and to provide a beam ~ 30 times brighter than the tungsten source. This will result in improvements in resolution of a SEM from ~ 50 Å to 30 Å as it improves the signal-to-noise ratio (Broers, 1970).

In addition, the higher brightness of this source makes it possible to use smaller apertures, so that only those electrons traveling very close to the axis of the beam are used in the image. This will minimize the spherical aberration effects in the magnetic lenses and improve the depth of field. It is a massive mechanically stable source (fewer vibration problems) and has a long life compared to that of the tungsten source. Thus SEM systems employing this type gun can be used in industrial fabrication and inspection applications (Black, 1971b; Broers and Hatzakis, 1972).

As shown in Fig. 1–9, the field emission source has a greater beam current than the LaB_6 source for small spot sizes, since its brightness is ~ 40 times that of the LaB_6 for equivalent spot sizes. An extremely strong electric field placed at the surface of the cathode suppresses the surface potential barrier (called the work function), which normally prevents the electrons from escaping, so that the electrons simply pour out of the tungsten cathode surface. These systems have very small apparent sources for the electrons even though the tungsten wire itself is $\sim 1,000$ Å in diameter. This is because when a voltage is applied, the lines of electric force emerge almost radially from the tip and thus guide the electrons on paths that appear to come from a virtual source of 100 Å or less in diameter inside the filament tip. This means, for example, that a crossover of 100 Å can be produced without employing any magnetic lenses. Unfortunately, this source requires ultrahigh vacuum conditions (10^{-10} torr) in the gun region, and even then the current stability is poorer than with heated cathodes. Systems employing the FE gun generally photograph the image off a TV screen while fast scanning for 10 to 40 sec to circumvent this problem.

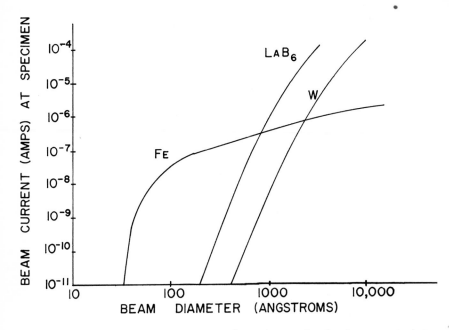

Fig. 1–9. Variations in beam current versus beam diameter for the three current electron gun sources used in SEMs. The LaB₆ source provides about one order of magnitude higher current than the tungsten source, but the field emission provides a higher current than either of the thermal types for spot diameter normally used in SEM work.

Table 1–1 compares the properties of standard electron guns. The LaB₆ gun can be installed on most of today's SEMs with minor modifications of the electronics and vacuum systems; some manufacturers already offer this gun as an option. While improving resolution, it will also allow for increased scanning rates for dynamic studies, so that rapidly changing processes can be examined dynamically in the SEM without loss of signal strength.

Resolution

It is generally assumed that topographical features smaller than the beam cannot be resolved by the SEM. The factors which affect resolution in the SEM can be placed in three groups:

> Group 1. Practical interference problems
> > (a) Sample contamination
> > (b) Mechanical vibrations
> > (c) Stray fields

Table 1–1   Electron Gun Characteristics for Tungsten, LaB₆, FE Sources

Characteristic / Cathode Type	Tungsten Hairpin "W" Heated	LaB₆ Pointed Heated	FE Field Emission (Tungsten)
Tip diameter	.005″	1 μm	.05 μm
Beam diameter at specimen (1 or more magnetic lenses)	50 Å	25 Å	10 or less Å
Beam current (amp)	8×10^{-13} to 10^{-12}	2×10^{-12}	1×10^{-10}
Beam brightness $\dfrac{\text{amp/cm}^2}{\text{steradian}}$	10^5	6×10^6	10 to 200×10^6
Operating temperatures °K	2859	2000	Ambient
Vacuum required	5×10^{-4}	1×10^{-6}	1×10^{-10} at gun; 10^{-6} in specimen chamber
Estimated life	30–40 hr at 20 kV	~3000 hr	Indefinite

Sample contamination can be most bothersome, as the source of the contamination can be quite varied. Inherent to all electron optical systems are beam contamination problems wherein the interaction of beam within the vacuum space deposits hydrocarbons on the interior walls of the SEM and on the specimen.

Contamination on the surface will retard the secondary emission and result in darkened regions in micrographs and a loss in resolution. Contamination can originate during specimen preparation or specimen aging (oxidation) in atmosphere. Nitrogen purging of the scope during specimen change helps reduce contamination problems and improves specimen change time. In general, group 1 problems should be eliminated by careful engineering by the manufacturer and skilled microscope practice by the user.

Group 2. Electron optical performance
    (a) Beam brightness
    (b) Lens aberrations
    (c) Lens operating conditions
    (d) Accelerating potential

The total beam diameter d or d_{min} is estimated by

$$d^2 = d_0{}^2 + d_s{}^2 + d_c{}^2 + d_f{}^2 \qquad (1\text{--}7)$$

where d_0 = aberrationless beam diameter at specimen

d_s = diameter of disk of confusion due to spherical aberration

$\quad = 1/2\ C_s\alpha^3$, where C_s is the spherical aberration constant.

d_c = diameter of disk of confusion due to chromatic aberration

$\quad = \dfrac{\Delta V}{V}\ C_c\alpha$, where C_c is the chromatic aberration constant,

V = acceleration voltage and ΔV the instability in that voltage.

d_f = diffraction effect or the Airy disk diameter, where

$$d_f = 2r = \frac{1.22\lambda}{\alpha}$$

The final beam current i was given by Eq. 1–5 as a function of d_0 and J_c, the beam current density at the cathode. The beam current density at the image plane is J_i, and is given by

$$J_i \simeq J_c \frac{eV}{kT}\,\alpha^2 \qquad (1\text{--}8)$$

where e = electron charge, V = accelerating voltage, k = Boltzmann's constant, T is absolute temperature, and α is the aperture angle as before.

To improve concentration of electrons on target, one can:

(1) Raise V (go to higher accelerating potentials).

(2) Increase arrival angle (use large apertures or shorter working distances).

(3) Select a cathode such as LaB_6, which will emit at lower temperatures.

(4) Select a cathode with higher specific emission (J_c) such as LaB_6 or FE.

All the above changes affect d. That d_0 is a critical parameter in the SEM will be explained below.

By rearranging Eq. 1–5, one can observe that

$$d_0{}^2 = \frac{6i}{\pi^2\beta}\,\alpha^{-2} \quad \text{where} \quad \beta = \frac{J_c eV}{\pi kT} \quad \text{or}$$

$$d_0 = C_0/\alpha \quad \text{where} \quad C_0 = \sqrt{\frac{6i}{\pi^2\beta}}$$

Thus $\qquad d^2 = \dfrac{C_0{}^2}{\alpha^2} + \dfrac{1.5\lambda^2}{\alpha^2} + C_s{}^2\alpha^6 + \left(\dfrac{C_c\Delta V\alpha}{V}\right)^2 \qquad (1\text{--}9)$

Thus there is a value of α for which there is a minimum value of d. This is shown in Fig. 1–10. For the SEM operating at 20 kV with $C_s = C_c = 1$ cm, $i = 10^{-11}$ amp, $\beta = 10^5$ amp/cm²/ster and $\frac{\Delta V}{V} = 10^{-4}$, d_0 and d_c are major terms and interset at $\alpha = .005$ radian where $d_0 = d_c = 50$ Å. Thus, $d = \sqrt{50^2 + 50^2} = 70$ Å, which is the maximum resolution typically achieved in the SEM. Notice that a 30-fold improvement in β by switching to a brighter gun will lower the value of d_0 and improve the resolution to about 35 Å. Loss of resolution beyond these levels can be due to wobble in the beam, astigmatism in the beam, vibrations in the stage or the column, and group 3 factors.

Group 3. Beam specimen interactions
(a) Electron beam penetration and scattering
(b) Secondary electron yield
(c) Specimen geometry
(d) Specimen material
(e) Specimen damage due to beam heating

Different specimens when examined with the SEM present widely varied beam-specimen interaction conditions because of variations in specimen material and geometry. One cannot be certain until actual examination what levels of resolution and detail will be observed. This problem will become more complex as new methods are developed to improve specimen preparation and secondary electron emission yield and/or collection. However, the resolution can be improved by increasing the brightness β as this will shift the d_0 set of curves to the left in Fig. 1–10. Increasing accelerating voltage will help because if, for example, we operated at 100 kV with $C_s \simeq C_c = 1$ mm, then d_s and d_f are most critical. In other words, $d_s \simeq d_f = 5$ Å with $d = 7$ Å, which is the typical resolution of the TEM or the FE STEM. Thus, d_o is not important in the TEM but is in the SEM, because in the latter it is necessary to restrict the beam penetration in the specimen so that voltages of 10 to 40 kV instead of 100 kV are employed. Lowering the kV increases the d_o, d_f, and to some extent d_c.

Because of its size, the specimen is located outside the magnetic field of the final lens in the SEM. This apparently places limits on the focal length of the lens and makes C_s and C_c of the order of 1 to 2 cm compared to ∼1 mm in the TEM. Thus the need to place the deflection coils above the final lens in order not to needlessly increase the focal lengths of the final lenses. The need for using magnetic lens, which have

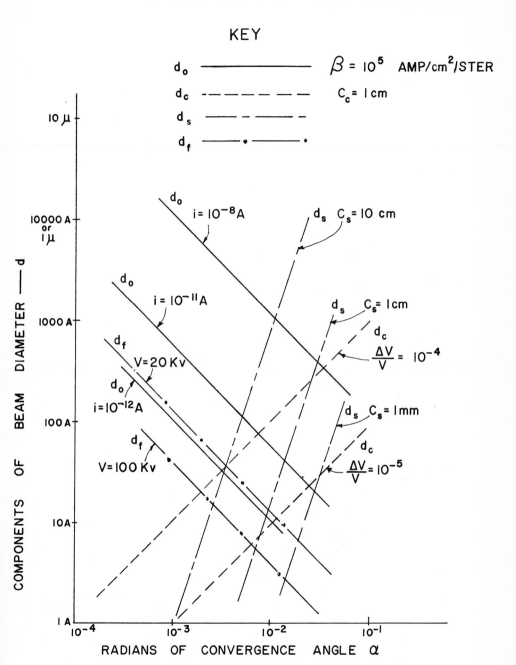

Fig. 1-10. Graphical presentation of the electron optical factors determining the effective spot diameter on the specimen as a function of the angular aperture for $\beta = 10^5$ amp/cm²/ster and $C_c = 1$ cm; $d_a =$ current carrying ability; $d_c =$ chromatic aberration effect; $d_s =$ spherical aberration effect; $d_f =$ diffraction effect.

lower spherical and chromatic constants than typical electrostatic lens, is also apparent. It is also clear that the beam brightness parameter holds the greatest promise for improvement in the resolution.

However, little further improvement in the resolution through reductions in probe diameter below ~25 Å for the SEM can be expected as the brightness of the electron gun is increased. Higher beam currents will improve the signal-to-noise ratio in the images as well as shorten exposure durations. In general, electron penetration effects rather than electron optical limits establish the ultimate resolution of the SEM (Broers, 1970).

Repeating the calculations with typical values for a FE SEM setup ($V = 20$ kV, $C_s = C_c = 1$ cm, $i = 10^{-10}$ amp, $B = 10^7$ amp/cm^2/ster., and $\Delta V/V = 10^{-4}$), it can be shown that d_c and d_0 are the major terms and $d \cong 20$ Å, the ultimate of resolution level which can be achieved in the SEM operated at these conditions.

Low-Voltage Operation

Most modern SEMs are designed to operate at accelerating voltages of 1 to 30 kV; many manufacturers offer a 50 kV option. Lowering the accelerating voltage to 5 or less kV results in less damage to the specimen due to beam heating. In Fig. 1–11, for instance, the edge of a glass knife, as used in microtomy, is examined with the SEM (Black, 1971a). Although the knife was gold-coated by vacuum evaporation, beam heating at 20 kV caused the coat to bubble at B, producing an image artifact. CE indicates the cutting edge of the knife with its associated striations or steps in rake face. Lower voltages also improve the visibility of the surface provided the resolution loss is not detrimental at the magnification desired. This increase in the visibility of surface details is due to the reduced penetration of the specimen by the probe.

Fig. 1–12 shows a series of micrographs of aluminum thin film chips lying on a TEM copper grid examined at different accelerating voltages. As the beam voltage increases, the electrons penetrate deeper into and through the specimen, which results in its becoming transparent. Notice that the grid bar structure can be observed under the chip at 30 kV. This phenomenon in a SEM is called double sources of secondaries, since secondaries are being produced at both the top where the beam enters and bottom where the beam exits surfaces at high kV's (Black, 1972).

Fig. 1–13 shows another similar study on an aluminum film at 20 kV and 3.3 kV, where again marked improvement in the surface visibility is obtained at the lower kV. At 1 kV, the depth of penetration, d_p, can be of the order of 100 to 500 Å, and the teardrop penetration configura-

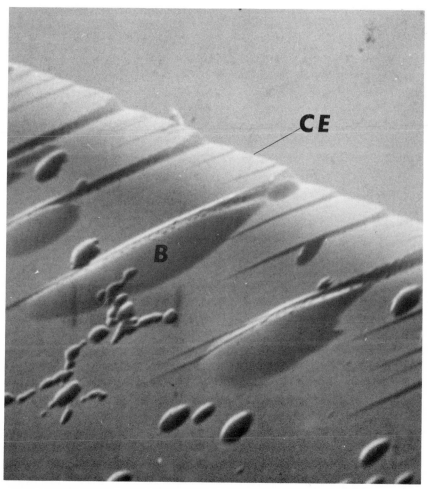

Fig. 1–11. Glass knife edge examined with the SEM at 20 kV. Gold coating has bubbled under beam heating at B. "CE" = cutting edge of knife.

tion (Fig. 1–4) balloons into a hemisphere and greatly increases the effective area of secondary emission. The effective wavelength of the probe electrons is increased according to Eq. 1–2, and thus diffraction effects will become more important. In fact, as one can observe in Fig. 1–10, both the spherical and chromatic aberrations are increased by decreasing voltage.

Thus an additional reason for using electromagnetic lenses rather than electrostatic lenses in the SEMs is to keep the aberration constants C_s

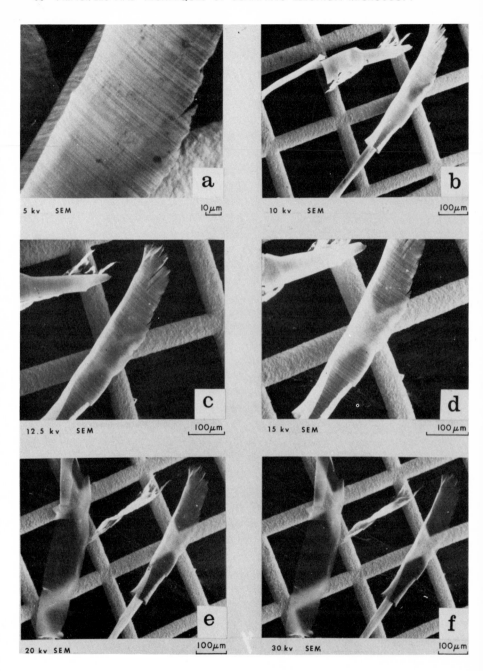

Fig. 1–12. Variation in surface visibility and transparency of thin aluminum film at different accelerating voltages.

and C_c as low as possible for all accelerating voltages. Independent of the absolute magnitude of C_s and C_c, there is an optimum value of alpha which produces a minimum value of spot size for a given current. Decreasing beam current decreases d_o, while decreasing voltage increases d_o. However, a decrease in the current is in conflict with noise problems, since one must produce a signal that will overcome the noise levels in the electron collection and amplification system.

Thus an effort is made to increase beam current density at the gun or source, because a decrease in the voltage will lower the beam current density at the image plane. In tungsten filaments, lower accelerating voltages (below ~10 kV) will result in significant impairments of the resolution. For some kinds of specimens, 1 kV may result in an increased efficiency in the secondary electron emission coefficient but a reduced efficiency in the generation of backscattered electrons. This may result in achieving higher resolutions.

A serious image defect common in SEM studies is specimen charging. An example of this is shown in Fig. 1–14, where the specimen was too complex in configuration for complete gold-coating, even when a rotary tilt stage was employed during evaporation. This problem can also occur from jarring the specimen after coating, which can break the continuity of the gold film. It has been shown that fast scanning at low voltages is an effective manner of examining charging or uncoated nonconductive materials (Oatley et al., 1965). When a high brightness gun is used, reasonable levels of resolution can be achieved.

Charging, however, can also be combated by using backscatter imaging, as the higher-energy electrons ignore the stray fields on the surface. In Fig. 1–15, polished metal surface of iron (Fe), silver (Ag), and carbon (C) has been examined. Note that the particle at D is charging in secondary mode and degrading the image, while in backscatter mode it is well imaged. The small particles lying on the iron surface are imaged better in backscatter, because the excessive secondary emission that they radiate, when struck by a 20 kV primary electron, is being omitted from the image. It should also be mentioned that the stray fields become significantly more serious disturbances to the imaging process at low voltages; however, it is possible to achieve resolutions of the order of 1,000 to 2,000 Å at 1 kV with properly designed W electron guns.

X-Ray Analysis

By using SEM, morphology of the surface can be revealed by every kind of signal except X rays and auger electrons. Element analysis or composition determinations, on the other hand, can be revealed by X rays,

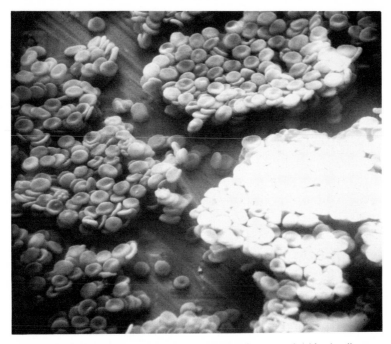

Fig. 1–14. A classic example of specimen charging in the SEM. The blood cells were prepared by standard chemical fixation and drying procedures and coated with gold. Carbon evaporation prior to gold or gold-palladium often eliminates charging problems in complex geometries.

cathodoluminescence, auger electrons, and backscattered electrons. There are basically two X-ray systems currently available to the SEM user; these are generally known as the wavelength dispersive method (WD), and the energy dispersive method (ED). Although these methods are used for the same purpose, they are quite different in form and characteristic results.

The conventional or traditional method of compositional analysis via X-rays has been the WD system in instruments known as electron probe microanalyzers (Fig. 1–16). In this system, X-ray spectrometers (metal crystals) are used to measure and identify and subsequently count X-rays emitted from the probe-surface interaction based on the wavelength of the emitted X ray. The energy of each X-ray photon corresponds to

Fig. 1–13. Thin aluminum film examined at 20 and 3.3 kV, demonstrating improved surface visibility at lower kV. Thin film was prepared by microtomy methods. Biological thin sections mounted on bare grids are very similar in topography.

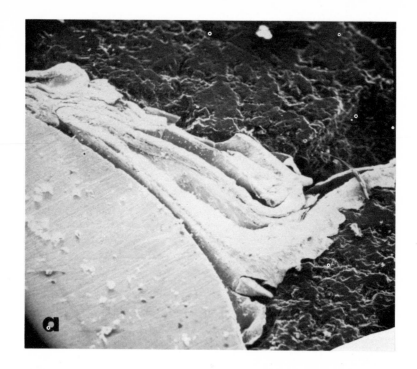

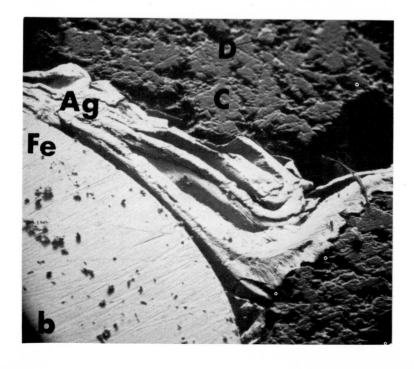

the traditional energy between orbital shells of the bombarded atoms according to the well-known expression

$$E = mc^2 \qquad (1\text{–}10)$$

where $E =$ energy, $m =$ mass, and $c =$ velocity of light.

The wavelength of the X-ray photons can be expressed in terms of Eq. 1–2, where momentum = mass times velocity = mv, and velocity is the speed of light or c. Then

$$\lambda = \frac{h}{p} = \frac{h}{mv} = \frac{h}{mc} \qquad (1\text{–}11)$$

and by combining these two expressions to eliminate mass one obtains

$$\lambda = \frac{hc}{E} \qquad (1\text{–}12)$$

wherein the wavelength is related to the energy. The phenomena of diffraction can be used to separate the wavelengths according to Braggs' law

$$n\lambda = 2d \sin \theta \qquad (1\text{–}13)$$

by means of an arrangement shown schematically in Fig. 1–16. In order to solve Eq. 1–13 mechanically, d is supplied by the analyzing crystal where d represents a particular set of crystal plane spacings; θ is the angles made by the crystal plane and the incident electrons, and is sought geometrically by physically shifting the position of the crystal and the detector while watching the X-ray counter for peak readings; n is the order of the reflection (has integer values).

The WD method has much better spectrum resolution and signal-to-background ratio than those shown by the ED method. Improvements in brightness and detection systems may result in improvements in low-current and high-resolution performance. In specimens for TEM, areas of a few hundred angstroms in diameter may be analyzed in the near future.

The energy dispersive system requires a detector whose output signal intensity is proportional to the energy of the incident quanta. In effect then, according to Eq. 1–12, the wavelength λ can be determined by measuring the equivalent X-ray quantum energy E. Detectors (e.g., gas flow proportional counter) capable of such measurements have been

---

Fig. 1–15. Polished surface of a series of metals examined by normal imaging methods in secondary mode (a) and in backscatter mode. (b) Backscatter was performed with secondary collector set to prevent admittance of secondary electrons.

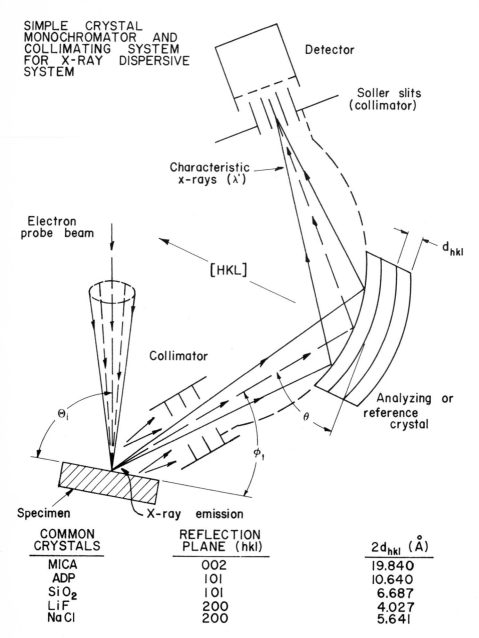

SIMPLE CRYSTAL MONOCHROMATOR AND COLLIMATING SYSTEM FOR X-RAY DISPERSIVE SYSTEM

Detector

Soller slits (collimator)

Characteristic x-rays (λ')

Electron probe beam

[HKL]

d_{hkl}

Collimator

Analyzing or reference crystal

Θ_i

θ

ϕ_1

Specimen

X-ray emission

COMMON CRYSTALS	REFLECTION PLANE (hkl)	$2d_{hkl}$ (Å)
MICA	002	19.840
ADP	101	10.640
SiO_2	101	6.687
LiF	200	4.027
NaCl	200	5.641

Fig. 1–16. Schematic arrangement of wavelength dispersive X-ray analysis system for the SEM or the electron probe microanalyzer.

available for some time; however, a better device is the lithium drifted silicon semiconductor detector.

X rays striking this material produce electron-hole pairs in proportion to the X-ray energy, which means that a current charge or pulse is produced in the silicon. This charge is fed to a field effect transistor (high gain, low noise) where it is amplified and fed to additional amplifiers and finally fed to multichannel analyzer after some shaping, as shown in Fig. 1–17. The voltage pulses are separated by amplitude and stored. The resulting spectrum can be displayed by various readout systems.

The ED systems are easily added to the SEM, and are much less complicated to use than WD systems. The former can be readily employed when the total X rays generated are weak, and can be used in thin film analysis as well. The system has some disadvantages compared to the WD system. Its resolution of spectra is very poor, and it is limited in detectability of alloys to those with $Z > 10$ or so.

The semiconductor silicon must be maintained at cryogenic temperatures, and must be isolated from contamination by a beryllium window which cuts out long wavelengths. However, its many advantages outweigh its disadvantages. The advantages include simple design and operation, speed of analysis, good detector resolution, and a high sensitivity. Because the collection efficiency is not sensitive to specimen height or defocusing effects during scanning, no X-ray focusing is necessary (a time-consuming job in a WD system), and completely automated, computer compatible systems are feasible.

A SEM with a backscatter detector and ED X-ray system is a versatile tool capable of extensive intricate analysis of both biological and non-biological specimens. Before purchasing an X-ray system, however, the reader should consult a paper by Beaman and Isasi (1971) on electron beam microanalysis. The application of such systems to the examination of biological thin sections for distribution of elements in the tissue structure is a remarkable advance in biological analysis. The distribution of Pb, Ca, and P in a kidney section of a lead-poisoned rat, for instance, has been presented by Carroll et al. (1970).

Stereopairs

An unusual feature of the SEM is the interpretation of the micrographs, since the contrast mechanism described for the SEM and that which operates in an optical image of the eye are analogous. Since we are accustomed to viewing TV, TV raster images, as produced by the SEM, are accepted as perfect reproductions of a microscaled surface. When possible, the mind supplies the third dimension in the image. However,

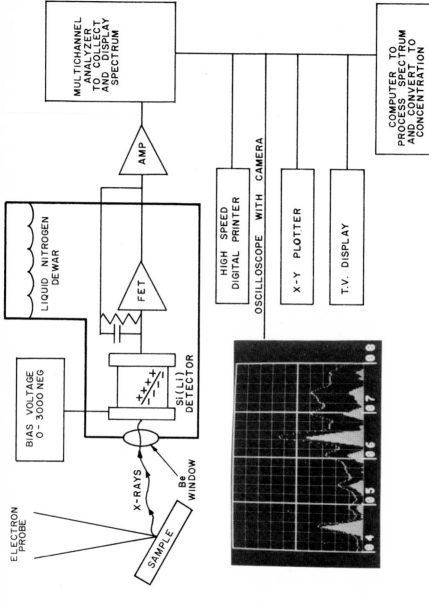

Fig. 1-17. Schematic arrangement for energy dispersive X-ray analysis system for the SEM or electron probe microanalyzer.

stereopairs of the microstructure are a safer alternative than a vivid imagination.

Although conventional micrographs obtained with the SEM have a three-dimensional appearance, stereopairs provide more accurate information on the microstructure of surfaces. One set of stereomicrographs can yield more information than can many conventional micrographs. Standard stereophotogrammetric techniques can be used to determine step or height differences in the microstructure. In addition to enhancing the interpretation of the microspace, stereopairs reveal image distortions produced by faulty specimen preparation or mounting, variations in the scan rates due to scan generator faults, specimen motion due to scan generator faults, specimen motion due to further drying or charging, and other imaging faults. This technique is applicable to backscatter, transmitted, cathodoluminescence, and secondary images.

The stereopair technique is especially helpful for viewing thick sections in transmission. This procedure permits the location (within the microstructure) of particles and the determination of height relationships of the often overlapping structures. If the signals are constructed on a point-to-point basis, the images can be stored and modified in computer hookups for later presentation. It seems feasible that this system can be combined with holography setups for truly three-dimensional presentation.

The conventional micrographs obtained with the SEM represents the view of the specimen as if the eyes of the viewer were looking down along the electron probe; the position of the collector is largely irrelevant. In order to prepare a stereopair, it is necessary to vary the angle between the incident beam and the specimen between two micrographs. The specimens are tilted to an angle of 6 to 12° between two micrograph exposures while maintaining the same area in the field of view and readjusting the focus (and perhaps exposure) for each exposure. The pair can be examined in a stereoviewer, and the structure analyzed in detail. Instead of tilting, one can rotate the specimen on certain axes to produce the stereo effect.

A stereopair of a watch mechanism is shown in Fig. 1–18. The left micrograph was obtained by tilting the specimen to a 14° angle, and the right one to 20°. Is the piece of debris (marked D in the left micrograph) attached to the gear or lying on the surface below the gear? Only stereoviewing of the pair can answer this question.

Boyde (1971) has demonstrated an anaglyph technique using Polaroid film. This technique eliminates the need for a stereoviewer but requires the replacement of the standard CRT with one having white phosphor. One micrograph is obtained in one color (e.g., using a green filter), and,

Fig. 1–18. Stereopair of watch mechanism obtained with the SEM with 6° tilt difference between micrographs. The piece of debris (D) is lying on the surface below the gear.

after tilting, a second one is obtained in a second color (using a red filter). The same two filters are then used to view the pair.

Practical Considerations

Image distortions can be produced by motion or drift of either the specimen or the beam during exposure. A very common problem for new users of the SEM is not to have the filament properly saturated and centered; an unsaturated filament can produce image instabilities. If undersaturated, the beam will not be stable, and the image will change focus during exposure. Such instabilities can also occur even for saturated filament owing to problems in the accelerating voltage or lens current supplies. Oversaturation of course needlessly shortens the life of the filament.

There is some confusion regarding procedures for obtaining proper saturation and alignment in various SEMs. Some instruments read beam current and/or filament current at the gun (as in the TEM), while others read the specimen current. Some columns are prealigned, while others allow for translation of the gun and first condensor lens. Saturation and alignment must be accomplished by reading meters and watching the CRT for image quality (whereas in the TEM one can watch the beam itself). Although it is not readily apparent to the user, SEMs require as much learning time to properly operate as do TEMs.

The user must also learn how to critically compensate the astigmatic conditions in the beam and align the final aperture. In some systems, where the aperture is withdrawn to isolate the specimen chamber during specimen change, aperture alignment should be constantly checked. Changing from one aperture to another almost always requires re-alignment of the aperture as well as astigmatic corrections. In the TEM, holey films are used to align and compensate the beam, but no such standard device has yet emerged for the SEM user. The ideal specimen would have edges which are well defined and in the same plane; a silver precipitate works well, as does a gold- palladium coated pure carbon holey film. The latter can be examined in the TEM for comparative pur-poses, and has simple, known overall geometry. Also, satisfactory stand-ards have not been developed yet for magnification and resolution de-terminations, the latter being of particular difficulty in that resolution is dependent upon the specimen-beam interactions as well as upon the electron optics.

The user should maintain a log book to keep a summary record of instrument performance and maintenance operations. This log book should include data on the following items:

(1) Dates, names, and times of use of the SEM, including beam time.

(2) Running number record of specimens examined and micrographs made on the instrument, including aperture and kV data.

(3) Dates and beam time when changes and/or cleaning of the fol-lowing have been performed:

(a) Filament, cathode, and anode (spare set recommended);

(b) Apertures (spare or new apertures recommended with proper cleaning);

(c) Scintillators (spares should always be available);

(d) Liquid N_2 in Dewars for X-ray units.

(4) Periodic (weekly) checks on resolution level should be performed and logged in the book. This is important, so that the level of perform-ance can be established for the instrument independent of the kind of specimen being examined. A quick check can be performed on resolution if the scope has a waveform monitor by the following simple steps.

(a) Examine a standard specimen which has a sharp edge or sharp changes in contrast (e.g., TEM coated holey grid).

(b) Set magnification at 100,000×.

(c) Place instrument in line scan mode with scan line in center of CRT frame.

(d) On wave form monitor, measure with cm scale the horizontal distance from a peak to a valley in the waveform. Then, for example, 1 cm peak-to-valley measured horizontally will be equivalent to ~1,000 Å resolution level in the SEM.

THE SCANNING TRANSMISSION ELECTRON MICROSCOPE (STEM)

The scanning transmission electron microscope is of recent development, and there are only a few commercial units as yet available. Many commercial SEMs can be used in the transmission mode by inserting a thin film in place of a solid surface and placing an electron detector system below the specimen (see Fig. 1–2). The same arrangement can be obtained in the TEM. The resolution in these hybrid STEM systems is not as yet as high as in the TEM. However, they offer the user increased specimen penetration, so that thicker specimens can be examined at the same level of kV. The STEM system devised by Crewe (1971) has greater resolution and image processing potential than the modified TEM or SEM systems. In this system, an electron sorting procedure allows the separation and collection of electrons into three groups (elastic, inelastic, and unscattered) simultaneously. Thus one can obtain a signal from each group, which is related to a point on the specimen. The resolution is equivalent to the TEM systems, but because the image is produced on a point-to-point basis, the signals can be stored, mixed, and modulated to produce a final image. The signals can also be computer-processed, since any scanning system lends itself to direct signal processing by computer. In terms of electron optics, a reciprocity can be shown to exist between the scanning transmission system and the TEM.

REFERENCES

von Ardenne, M. (1938). The scanning electron microscope: Practical considerations. *J. Phys.* **109**, 533.

Baxter, A. S. (1949). Ph.D. Dissertation. University of Cambridge, Cambridge, England.

Beaman, D. R., and Isasi, J. A. (1971). Electron beam microanalysis. Parts I and II. *Materials Res. & Stds.*, **11**, Nos. 11 and 12.

Black, J. T. (1971a). Ultramicrotomy of embedding plastics. *Applied Polymer Symposium*, No. 16, p. 105. Wiley and Sons, Inc., New York.

Black, J. T. (1971b). SEM nondestructive testing: Applications and potential. *Int. J. Non Destruct. Test.* **3**, 1.

Black, J. T. (1972). Thin film surface morphology visibility. *Proc. 30th EMSA Meeting*, p. 402.

Boyde, A. (1971). A review of problems of interpretation of the SEM image with special regard to methods of specimen preparation: Scanning electron microscopy. *Proc. 4th Ann. Symp.*, I.I.T. Research Institute, Chicago.

Brachet, C. (1946). Note on the resolution of the scanning electron microscope. *Bull. Assoc. Tech. Maritime Aerom.* **45**, 369.

Broers, A. N. (1970). Factors affecting resolution in the SEM: Scanning electron microscopy. *Proc. 3rd Ann. Symp.*, p. 1. I.I.T. Research Institute, Chicago.

Broers, A. N., and Hatzakis, M. (1972). Microcircuits by electron beam. *Scient. Amer.* November issue, p. 34.

de Broglie, L. (1924). A tentative theory of light quanta. *Phil. Mag.* **47**, 446.

Busch, H. (1927). The calculation of the paths of cathode rays in an axially symmetric electromagnetic field. *Ann. d. Physik.* **81**, 974.

Carroll, K. G., Spinelli, F. R., and Goyer, R. A. (1970). Electron probe microanalyzer localization of lead in kidney tissue of poisoned rats. *Nature* **227**, 1056.

Crewe, A. V. (1971). A high resolution scanning electron microscope. *Scient. Amer.* **224**, 26.

Crewe, A. V., Eggenberger, D. N., Wall, J., and Welter, L. (1968). Electron gun using field emission source. *Rev. Sci. Inst.* **39**, 576.

Davoine, F. (1957). Secondary electron emission of metals under mechanical strain. Ph.D. dissertation, University of Lyons, France.

Davoine, F., Bernard, R., and Pinard, P. (1960). Fluorescence of alkali halides observed with a SEM. *Europ. Reg. Conf. Electron Micros.* Delft, Netherlands, p. 165.

Everhart, T. E., and Hayes, T. L. (1972). The scanning electron microscope. *Scient. Amer.* **226**, 54.

Everhart, T. E., and Thornley, R. M. F. (1960). Wide band detector for microampere low-energy electron currents. *J. Sci. Inst.* **37**, 246.

Grivet, P. (1965). *Electron Optics.* Pergamon Press, London.

Heidenreich, R. D. (1964). *Fundamentals of Transmission Electron Microscopy.* John Wiley and Sons, New York.

Jenkins, F. A., and White, H. E. (1951). *Fundamentals of Optics.* McGraw-Hill, New York.

Kimoto, S. (1972). The scanning microscope as a system. *JEOL News* **10e**, 1.

Knoll, M., and Ruska, E. (1932). Geometric electron optics. Parts I and II. *Ann. d. Physik.* **12**, 607.

Knoll, M. (1935). Static potential and secondary emission of bodies under electron irradiation. *Tech. Phys.* **16**, 467.

McMullen, D. (1953). Investigations relating to the design of electron microscopes. Ph.D. Dissertation, Cambridge University, England.

Oatley, C. W., Nixon, W. C., and Pease, R. F. W. (1965). Scanning electron microscopy. *Adv. in Electronics & Electron Physics.* **21**, 181.

Pease, R. F. W. (1963). High resolution scanning electron microscopy. Ph.D. Dissertation, Cambridge University, England.

Smith, K. C. A., and Oatley, C. W. (1955). The scanning electron microscope and its fields of application. *Brit. J. Appl. Phys.* **6**, 391.

Thompson, J. J. (1897). Cathode Rays. *Phil. Mag.* **44**, 293.

Thornton, P. R. (1968). *Scanning Electron Microscope.* Chapman & Hall, Ltd., London.

Zworykin, V. K., Hillier, J., and Snyder, R. L. (1942). A scanning electron microscope. *ASTM Bulletin* **117**, 15.

Zworykin, V. K., Morton, G. A., Ramberg, E. G., Hillier, J., and Vance, A. W. (1945). *Electron Optics and the Electron Microscope.* John Wiley and Sons, New York.

2. CRITICAL POINT DRYING

Arthur L. Cohen

Electron Microscope Center, Washington State University
Pullman, Washington

INTRODUCTION

The transmission electron microscope (TEM) was designed for specimens of some electron transparency. Yet almost as soon as it was commercially available, replicating and shadowing methods were devised for the examination of surfaces and to provide three-dimensional view (Bradley, 1965). Biologists were aware that some distortion was produced when soft cells were dried; nevertheless, most were content with the highly detailed images of dried and shadowed viruses, bacteria, and other small organisms.

The commercial advent of the scanning electron microscope (SEM) in the mid 1960s stimulated the nearly dormant demand for specimens with undistorted surfaces. The SEM is eminently suited for examining the surfaces in detail and with impressive depth of focus. The biologist could not remain content with the wrinkled, shriveled, and flattened specimens whose every artifact of drying was glaringly evident.

There are many types of specimens which, at least on the surface, are not affected significantly by drying. These include hard or hard-coated specimens such as: bone, shell, diatom, and protozoan skeletons; wood; the chitinous covering of insects and other arthropods; pollen grains; and the spores of some fungi and other plants. However, a hard covering is not necessarily a complete armor against collapse and distortion. Hard outer coatings must be sufficiently rigid to withstand the stresses caused by the drying and shrinking protoplasm. Most vegetative plant cells, pollens, spores of lower plants, and small arthropods shrivel when air-dried (Figs. 2–6, 2–8, and 2–10). Collapse and distortion are essentially total in the case of soft cells and tissues. Without some treatment to counter the effects of simple drying, many specimens are not suitable for examination with the SEM.

Causes of Distortion

When a soft specimen (e.g., a cell) dries, it usually shrinks. If the specimen is an elongated one, it twists; if cylindrical, it collapses. The original smooth surface may contract and tear, or it may wrinkle severely. These distortions caused by dehydration may be conveniently divided into two types—volume changes, and surface changes.

Volume Changes

The simplest case of volume change occurs when fluid under pressure is removed by rupture, evaporation, or simply by osmosis into a hypertonic surrounding medium. Mature living plant cells, with their large central vacuole filled with liquid under pressure, show such shrinkage strikingly, and are classic objects for demonstrating plasmolysis. It is well known that during dehydration through increasing concentrations of ethanol or other dehydration fluids, the exchange of liquids must be carried out in small steps; otherwise, water will rush out faster than the exchange liquid can enter, and the cells will remain in a permanently shrunken condition.

Water has a structural role in the macromolecular architecture of cell walls and protoplasm. These structures maintain their shape, in part, by the water molecules intercalated between the macromolecular network which forms their skeleton. As the water molecules are withdrawn without either replacement or other means to keep the skeletal meshwork locked in place, the molecules of the framework are drawn together, and the whole structure shrinks. In relatively large cells, the volume/surface ratio is great. Therefore, volume stresses are dominant. As cells become smaller, the volume/surface ratio decreases and surface forces predominate. A liter of water needs a fairly rigid container to hold it against gravity, but a small drop of water maintains its shape with nothing more than its own surface tension.

Surface Changes on Drying

Although volume stresses become less important with decreasing size of the specimen, other deforming stresses increase. At cellular and subcellular levels, the most severe stress is surface tension or, more precisely, interfacial tension. Pure water has the high interfacial (surface) tension of 73 dynes/cm against air at ordinary temperatures.

Volume stresses are at right angles to the surface. They may be compared to the forces which try to punch a hole through a membrane. Surface forces are parallel to the surface, and stretch or contract a membrane laterally. Since the surface proportionately increases as the volume decreases, these forces become greater with smaller volume. When cells are allowed to dry on a substrate, the receding water meniscus passes along and through them. As a result, enormous stresses may be set up. Anderson (1956 and 1966) has made the well-known calculation that the stress through a bacterium flagellum drying across a gap is 46,000 kg/cm^2.

Countering Volume and Surface Stresses during Drying

There are several procedures by which the stresses may be lessened or practically abolished.

Maintaining the Specimen Moist. For transmission microscopy, this procedure is feasible only at high or very high voltages, and is currently the object, but not the tool, of investigation. Minute specimens (e.g., bacteria) are maintained in a blister or environmental cell (Lane, 1970a; Swift and Brown, 1970; Fullam, 1972). The resolution is degraded because of the increased scattering. In the SEM, while there is a growing practice of maintaining specimens moist in an environmental chamber or observing them before they are dried, the preparation must be considered temporary, and examination must be completed before the specimen begins to distort, which usually occurs in a few minutes. Wet specimens have the advantage in that charging is minimized since they are electrically conducting, and therefore coating is not needed. Specimens may be maintained more permanently moist by substituting less volatile glycerin or glycols for the water (Heslop-Harrison, 1970; Idle, 1971; Mozingo *et al.*, 1970; Panessa and Gennaro, 1972a,b, 1973; Richter *et al.*, 1969). However, it appears that surface detail is sometimes obscured by a liquid film. Contamination of the microscope by glycerol is another drawback (Panessa and Gennaro, 1973, p. 402). The method must be used with caution.

Specimen Hardening by Fixation. Osmium tetroxide, mercuric bichloride, and aldehydes have been used to harden specimens so that they can better withstand the stresses of air-drying immediately or after dehydration with other fluids. As fixation is essentially a routine procedure for SEM preparation, only selected comprehensive reviews and studies are cited here (Arnold *et al.*, 1971; Bessis and Weed, 1972; Boyde, 1972; Boyde and Vesely, 1972; Boyde and Wood, 1969; Landboe-Christiansen and Parapat, 1972; Panessa and Gennaro, 1972a; Sträuli and Haemmerli, 1971).

Dehydration with Fluids of Low Surface Tension. Dehydration usually after fixation, through increasing concentrations of substituent liquids (e.g., alcohol, acetone, and propylene oxide) followed by air-drying is a popular method for alleviating the distortions caused by drying (Bessis and Weed, 1972; Boyde, 1972; Boyde and Vesely, 1972; Boyde and Wood, 1969; Nowell *et al.*, 1972; Shimamura and Tokunaga, 1970). For many tissues this procedure has been quite successful, but some published micrographs show from minor to severe distortion. The usefulness of fluid-substitution dehydration is probably due to the fact that the interfacial tension of these dehydration liquids is considerably less than that of water, and that they are fixatives as well as dehydration agents and hence harden the specimens against stress.

Freeze-Sublimation. Fresh or fixed specimens are quickly frozen by plunging them into liquid nitrogen or some other quenching fluid (e.g., Freon 22) at or near liquid nitrogen temperature. The quick freezing at very low temperatures reduces the size of the ice crystals and thereby the scale of disruption caused by them. While still frozen, the specimens are placed in a vacuum, and the water sublimes until all the ice has evaporated. The freeze-sublimation (also called freeze-drying) method can yield excellent results (Boyde, 1972; Boyde and Wood, 1969; Landboe-Christiansen and Parapat, 1972; Marszalek and Small, 1969; Pasternak *et al.*, 1970; Paulin and Bussey, 1971; Small and Marszalek, 1969; Small and Ranganathan, 1970). Comprehensive discussions of theory (MacKenzie, 1972) and practice (Rebhun, 1972) of freeze-drying have been presented. The conditions for the SEM are not as stringent as those for the TEM.

It is also possible to combine certain advantages of freezing and observation of fresh material. Specimens may be frozen, placed in the SEM, observed during sublimation of the ice, and photographed when surfaces are free of ice but before drying causes distortion (Echlin *et al.*, 1970; Nei *et al.*, 1972, and in this volume).

The advantages of freeze-sublimation are: (a) relative simplicity of procedure; (b) apparatus may be made up from common laboratory components, or an evaporator may be used; and (c) unlike the critical point method, high pressures are not involved. The disadvantages are: (a) freeze-dehydration is a slow procedure (24 hr or more); (b) it can tie up rather expensive equipment, such as evaporators, and contaminate pump oils with water; (c) not all laboratories at present have liquid nitrogen or other cryogenic sources available; (d) quenching fluids are dangerous, and can cause severe frost burn; (e) specimens must be quite small for effective preservation of the fine structure by rapid freezing and for drying to be accomplished in a reasonable period of time; and (f) some specimens show severe surface wrinkling and other distortion. Several workers, by comparing freeze-drying with critical point drying, have concluded that generally critical point drying (CPD) is superior (Anderson, 1956, 1966; Bibel and Lawson, 1972; Cohen *et al.*, 1968; Falk *et al.*, 1971; Nemanic and Pitelka, 1971; Nowell *et al.*, 1972).

Critical Point Drying. This is the only method besides freeze-sublimation which essentially avoids surface tension effects. As it is the subject of this chapter, it is discussed in detail in the following sections.

The following discussion emphasizes some subjects which have been neglected, misunderstood, or taken for granted, and which are nevertheless sources of bewilderment and trouble, particularly for the newcomer to the field. Containers for specimens, and specimen mounting for examination, are two such topics.

THE CRITICAL POINT DRYING METHOD

If a liquid is placed in a sealed container, leaving some vapor space around it, and the container heated, the vapor pressure and hence the vapor density become greater. At the same time, the liquid (whose volume change is a function of temperature and little affected by pressure) expands and becomes less dense. There comes a point, the critical point (CP), a combination of pressure and temperature, at which the density of the vapor is the same as that of the liquid, the critical density (D_c). At this point, the surface tension is zero, and there is no difference between liquid and gas; the interface between them becomes flatter, fainter, and finally vanishes. The pressure at which the distinction between liquid and vapor disappears is the critical pressure (P_c); the temperature is the critical temperature (T_c).

The critical temperature has another property: above it, no amount of pressure will condense the gas into a liquid. Below it, liquid and vapor can coexist if the pressure is sufficient (greater than the vapor pressure). These conditions are shown in Fig. 2–1. If an object is immersed in the liquid below the critical temperature, and the system carried through the critical point to a temperature above critical temperature, the specimen is now in a gas phase without ever having passed through the liquid surface. By keeping the temperature above critical temperature, the valve can be opened to the air, and the pressure allowed to drop to atmospheric. The specimen has now been critical point dried. The critical point should be more aptly named the critical region, as is shown in Figs. 2–1a and b, where the critical points are loci on a smoothly changing curve, and not the sharp cusps inferred by the term "point."

At a high temperature (T_h), well above the critical point, the curve is a smooth hyperbola fairly closely following the inverse relation of temperature and pressure expressed by the classic gas law, $PV = nRT$. At a temperature (T_{c+}) just above the critical point, there is an inflection at the critical pressure. At a temperature (T_{c-}) below the critical temperature, the extreme right-hand part of the curve shows the fluid to be totally in the gaseous state. On compressing, it condenses to liquid, and the horizontal portion shows that volume is essentially independent of pressure. A slight increase in pressure condenses more of the vapor to an essentially incompressible liquid; slightly less pressure allows evaporation until equilibrium between liquid and vapor is restored. At sufficiently

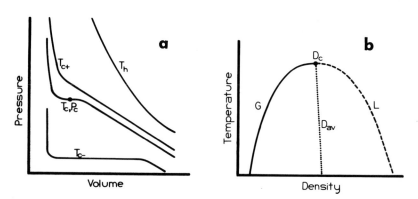

Fig. 2–1. (a) Relationship between pressure, volume, and temperature of a fluid near the critical point (see text for explanation). (b) Relationship between temperature and density of the liquid and gas phases of a confined fluid at the critical pressure. With rising temperature, the density of the liquid (dashed line L) decreases and the density of the gas (solid line G) increases. Their meeting point is the critical density D_c. The dotted line D_{av} illustrates the law of rectilinear diameters, which states that the average of the densities of a liquid and vapor at the critical pressure forms a straight line with temperature.

high pressures, the curve (left) becomes vertical; volume is independent of pressure. All vapor is condensed to liquid, and the vertical portion of T_{c-} (left) represents the practically incompressible liquid.

The formal mathematical description of a fluid at its critical point is forbiddingly complex; however, a qualitative description can be given. At the critical point, gas molecules are under sufficiently great pressure but at a low enough temperature so that they can coalesce to denser "clots" without sharp boundaries. These clots are sufficiently large to scatter light; hence at the critical point, particularly if approached from the high temperature side, a hazy opalescence develops, the critical opalescence. A slight lowering of temperature allows the rapid coalescence of these regions to form a liquid phase with a definite boundary.

So delicately poised are the forces of thermal energy, which tend to keep the molecules moving freely against the attractive forces, which tend to make them coalesce, that in the critical region very slight gradients of temperature or other perturbations are highly magnified. The fluid does not appear homogenous, but has the appearance of streams of different densities in constant motion. If another fluid, such as Freon TF, miscible with Freon 13 is present, the denser component is at the bottom. If the chamber is tipped, the bottom layer can be seen to surge back and forth, although there is no obvious boundary with the upper gaseous phase. The tendency of dense intermediate fluids to concentrate at the bottom has a practical advantage: the bomb can be partly drained from the bottom, leaving the specimens immersed in the pure fluid which will complete the transition to the gas phase.

The region without boundary between liquid and gas and the mathematically defined critical point do not exactly coincide. The boundary disappears at a slightly lower temperature than does the critical point, determined by other criteria (zero surface tension, uniform density) (Kimball, 1951). Discussions of critical phenomena have been presented by Blinder (1969), Rowlinson (1968), and Widom (1967).

Properties of Fluids

The fluids which undergo phase change from liquid to gas at the critical point are called the transitional fluids (Cohen *et al.*, 1968). Water is not completely miscible with any of these transitional fluids in their liquid state with the questionable exception of nitrous oxide (Boyde, 1972; Koller and Bernhard, 1964; Nemanic, 1972). Therefore a water-miscible dehydration fluid (e.g., ethanol or acetone) is used in steps of increasing concentrations to gradually replace the water. If the pure dehydration fluid is not completely miscible with the transition fluid, an intermediate fluid miscible with both may be interposed. Sometimes it is advantageous, as shown later, to interpose an intermediate fluid even when the pure dehydration fluid is completely miscible with the transition fluid. In a strict sense, any fluid interposed between the original aqueous environment of the tissues and the transitional fluid is an intermediate fluid. The term "intermediate fluid" (without distinction between dehydration and intermediate functions) is often used in the literature in this sense with little danger of confusion.

If its critical properties were appropriate, water would obviously be the ideal transitional fluid for critical point drying. However, the properties of water are far from ideal since it has $374°$ T_c and 217.7 atm (3,184 lb/in^2) P_c. There are chemicals with considerably more reasonable critical pressures and temperatures. Unfortunately, most of these are highly toxic, corrosive, or dangerously inflammable. The choice is limited; the fluids which have been used for the critical point drying method are shown in Table 2–1.

Carbon Dioxide. Carbon dioxide was introduced by Anderson (1951, 1956, and 1966) in order to maintain the three-dimensional shape of bacteria for examining with the TEM. Anderson stated that liquid carbon dioxide is miscible with amyl alcohol, but not with ethanol or water.

Table 2–1   Formulas, Names, and Critical Constants of Fluids Used in Critical Point Drying[1]

Name	Formula	Critical Temp. °C	Critical Pressure lb/in²	atm
Carbon dioxide	CO_2	31.1	1073	72.9
Nitrous oxide (Laughing gas)	N_2O	36.5	1054	71.7
Freon 13	$CClF_3$	28.9	561	38.2
Freon 23	CHF_3	25.9	701	47.7
Freon 116[2]	CF_3-CF_3	19.7	432	29.4
Freon TF (Freon 113)	CCl_2F-$CClF_2$	214.1	495	33.7

[1] Data on CO_2 and N_2O were obtained from *Handbook of Physics and Chemistry* (50th ed., 1969); E. I. Du Pont de Nemours and Co. *Technical Bulletin B-2* (1971 edition) was the source of data on the fluorocarbons (Freons). There are minor variations in data from different sources.

[2] The critical constants for Freon 116 are correct according to communication from the Du Pont Company. Earlier editions of *Technical Bulletin B-2* gave 24.3°C for the critical temperature and 480 lb/in^2 (32.6 atm) for the critical pressure. Curves and tables do not necessarily agree, even in the same edition.

Amyl acetate is miscible with ethanol, but not with water. Therefore the complete substitution series is: water/ethanol/amyl acetate/carbon dioxide. However, as discussed later, the amyl acetate step may be eliminated (DeBault, 1973).

Bacteria mounted on a Formvar film on grids were dehydrated through increasing concentrations of ethanol dehydration fluid, passed through amyl acetate as an intermediate fluid, and then placed in the simple bomb designed for critical point work. The bomb, attached to a tank of carbon dioxide, was cooled by raising a bucket of cold water around it; the carbon dioxide was admitted by a valve, and an outlet valve was slightly opened to flush out the amyl acetate. When the bomb was filled with essentially pure liquid carbon dioxide, it was immersed in a bucket of warm water, and the carbon dioxide brought above its critical temperature. While still kept warm, the outlet valve was slightly opened, and the pressure allowed to drop gradually to atmospheric level. Anderson (1956, 1966) gives detailed directions for the preparation of specimens on grids and subsequent treatments.

The primary purpose of the method is to retain the three-dimensional structure of a specimen. Stereomicrographs are useful for demonstrating the structure; Anderson provides detailed directions for their preparation.

Nitrous Oxide (N_2O). Koller and Bernhard (1962) described the use of nitrous oxide as a means of preparing tissues in an uncollapsed state without solvents or an embedding medium for ultramicrotomy. They selected nitrous oxide, presumably completely miscible with water, as a combined dehydration and transitional fluid. Koller and Bernhard used an elaborate bomb which contained a stirrer for circulating the liquid nitrous oxide (under high pressure) around the specimens and which incorporated a drying agent to remove the water from the circulation liquid. The process was completed in 24 hr. These authors pointed out that it was not absolutely necessary to go through the critical point; if the gases were discharged near the critical pressure and the critical temperature, the surface tension would be negligible. The last assumption may be questionable. Nitrous oxide appears to be more the subject of experiment than the means of routine critical point drying, as it is not apparently completely water-miscible (Boyde, 1972; Nemanic, 1972).

The Fluorocarbons (*Freons*). Cohen *et al.* (1968) pointed out that several of the fluorocarbons used as refrigerants, commonly known as Freons (also Genetrons, Arctons), have desirable properties as transitional fluids (Table 2–1). The critical temperatures are near room temperature, and the critical pressures are little more than half those of carbon dioxide or nitrous oxide. These lower pressures are in the range of inexpensive commercial refrigeration components, and it was easily possible to monitor the process through a viewing port.

One of the fluorocarbons, Freon TF (also known as Freon 113; CCl_2F-$CClF_2$), is liquid at room temperature, is nearly nontoxic, noninflammable, and commercially available as a solvent and cleaning agent of high purity. Because its chemical composition is similar to that of a transitional fluid, it seemed likely that all traces would not have to be flushed out (Cohen et al., 1968), as is the case with the other critical point drying methods, which require thorough removal of the intermediate fluid. A small amount of Freon TF carried over into the bomb forms a mixture, and acts like a pure transitional intermediate fluid with elevated critical pressure and critical temperature—in other words, with critical properties proportional to the relative concentrations. Hence, the prolonged flushing of the bomb was avoided, and there was increased assurance that no liquid would be left in the specimen to undo the critical point drying after pressure was released and the specimen was brought back to room temperature. This indeed proved to be the case, and although ethanol will mix with any of the Freons listed, Freon TF is used commonly as an intermediate to insure complete conversion to the gas and to save flushing with these rather expensive transitional fluids. The use of Freon TF brings the cost of the Freon critical point method down to approximately the same level as that of carbon dioxide.

Toxicity of Intermediate and Transitional Fluids

The toxicity of the transitional fluids (N_2O, CO_2, Freons) has been the subject of some concern and of considerable exaggeration in the cases of N_2O and the Freons. There has been little consideration of the much greater toxicity of some intermediate fluids which are emitted into the atmosphere when a bomb is flushed.

Nitrous oxide has been called extremely toxic (Nemanic, 1972). This is probably due to confusion with the other oxides of nitrogen, which are quite poisonous. However, nitrous oxide is the "laughing gas" of dental anesthesia, and is used as a propellant for the instant whipped cream and other items sold in pressurized cans and intended for human consumption.

A high toxicity has been claimed for the Freons (IITRI discussion, 1972), although they seem to possess properties of inert gases. Transitional fluids listed in Table 2–1 have estimated toxicities somewhat lower then carbon dioxide (Du Pont, 1971). Freons used as propellants in aerosol cans have been inhaled in very high concentrations for the psychological effects. Under such conditions they are indeed dangerous, by asphyxia (oxygen displacement) and by freezing the vocal cords (Deichman and Gerarde, 1969) and partly by causing cardiac irregularities. The fluorocarbons can be converted to extremely toxic halogenated carbonyls by exposing them to high temperatures. For this reason open flames or very hot metal surfaces (i.e., hot plates at or near redness) should not be near the discharge port of a bomb.

The Freons (Du Pont, 1971; Lester and Greenberg, 1950) and nitrous oxide (Price and Dripps, 1965) must displace oxygen to exert toxic effects; carbon dioxide may be fatal at a concentration of 10%, and hence is the most toxic of the transitional fluids (Wollmann and Dripps, 1965). Despite the low toxicity of transitional fluids, critical point drying should be carried out with ample ventilation, more because of the possible noxiousness of intermediate fluids (which are discharged as the bomb is flushed out) than of the transitional fluids.

Ethanol, acetone, amyl acetate, and Freon TF are the commonly used intermediate fluids. Of these, amyl acetate (isoamyl acetate) (Fassett, 1963) and acetone (Rowe and Wolf, 1963; Deichman and Gerarde, 1969) are the most toxic, and prolonged breathing of their vapors may cause headache and other intoxication symptoms. Amyl acetate, which is particularly bothersome, should always be flushed into a hood. Mice exposed to 3% ethanol for 1 hr were killed (Treon, 1963). Freon TF has very slight toxicity (Du Pont, 1971), probably only little more than carbon dioxide. However, Freon TF may temporarily increase sensitivity to adrenalin and predispose to irregular heart beat (Clayton, 1967). This discussion has been based on acute toxicity, as the Freons seem to have no chronic effects (Gleason et al., 1969). Nevertheless, frequent, recurrent, and prolonged use should be safeguarded by thorough ventilation, as it should for any solvent.

The Bomb

Essentially the bomb is simple. It consists of a sturdy container with a sealable port for introducing the specimens, valves for admitting the transitional fluid and for letting its vapor escape, a pressure gauge, and connection to the supply tank. There must be some means of heating and cooling the chamber; the simplest method is to have all connections rigid and so arranged that the vessel may be immersed by raising and lowering containers of water of appropriate temperatures around it. This was essentially Anderson's original (1951) carbon dioxide bomb, and it remains the basic design for several commercial models.

Safety Precautions

For accurate monitoring of the process and, more important, for reasons of safety, a pressure gauge, in communication with the bomb at all times, is absolutely essential. The gauge should read approximately twice the intended maximum pressure, psig (pounds per square inch gauge against pressure of surrounding atmosphere). For Freons, the scale should extend to 2,000 psig; for carbon dioxide, 3,000 psig. The experimenter who designs his own apparatus should always remember that he is working with dangerous pressures. All units subject to pressure should have a safety factor of threefold or better. If the bomb is heated by running hot water, the temperature should never exceed 65°C. If it is heated electrically, the heater should not be allowed to rise above the same temperature. This author modified small hot plate thermostate switches to make them open at 65° at the highest setting.

A rupture disk is a desirable investment for the sake of safety. If a viewing port is used, it should be known to withstand the pressure; under no circumstances should a chipped or scratched glass be used. It is advisable to have a Plexiglas or other transparent shield between the operator and the bomb. The bomb should not be subjected to temperatures considerably below 0°C. Liquid nitrogen or dry ice temperatures should not be used without expert advice on the properties of the materials under pressure at these temperatures. At very low temperatures, metals become brittle, plastic and rubber seals fail, and operation is uncertain and unsafe.

A small, high-pressure jet stream from a crack, although not as devastating as an outright explosion, can cut like a knife. Sealing of connections should be carried out with appropriate gaskets and Teflon sealing tape, and never with any sealing compound which could contaminate the specimens. Although there are several excellent and tested critical point drying devices on the market, described in the following sections, the experimenter who nevertheless plans to build such an apparatus should obtain the advice of someone familiar with pressure chambers.

Other Design Considerations

Any critical point drying device, commercial or homemade, should be designed to cycle through the temperatures at a controllable and convenient rate. Walls sufficiently thick to withstand the internal pressure also help to insure even and gradual heating. Heating should be gradual (at least 3 to 5 min for a 15 ml capacity bomb) to allow the contents to remain close to thermal equilibrium with the walls. Too rapid heating with a large temperature gradient allows the liquid to boil violently at the hottest wall while condensing elsewhere; the strong currents can dislodge and damage specimens. If electrical heating is employed, the smallest hot plate or heating element capable of raising the temperature to 15° above the critical point should be used. A more massive element has too great a heat capacity, and can continue to heat the bomb to dangerous levels after it has been turned off.

Low temperatures (i.e., 5 to 15°C) are preferable for loading and filling the bomb. The intermediate fluid is less likely to evaporate completely at a low temperature before the vessel is closed, and the bomb fills more rapidly with the transitional fluid. A low tank pressure is necessary to fill it. The only limitation to starting the cycle near 0°C is set by humidity. Any moisture condensed in the bomb can ruin the specimens.

A thermometer either within the bomb or in the surrounding medium is also necessary. It should have a rapid response time; for this purpose, either a thermistor probe may be sealed into the bomb cavity, or, if the wall of the bomb is sufficiently thick, the stem of a dial thermometer may be inserted into a hole and sealed with thermally conducting grease. The thermometer of an original bomb design (Fig. 2–3c) (Cohen *et al.,* 1968) is too slow in responding to temperature changes.

If the bomb is designed to use only the Freons, components are relatively inexpensive and easily obtained from refrigeration supply firms (Cohen *et al.,* 1968). The increased pressure of the carbon dioxide-critical point drying method requires special, costly fittings.

A viewing port, especially if the Freons are used, is a useful adjunct. The bomb may be filled and flushed under observation, and the critical point phase transformation can be observed. An observation port shows if some intermediate fluid remains or if heating is insufficient. A port helps in flushing, for the liquid may be nearly completely flushed out and the chamber refilled. Otherwise a lengthy, continuous flush is required, which is time-consuming and also expensive if the Freons are used. The port is very useful for monitoring the supply condition of the reservoir. One becomes quite practiced at estimating the amount of transitional fluid left in the supply by the rate it enters the bomb. A port saves time and material by letting the observer see when the specimens are fully covered and the bomb is at least three-fourths full. It is neither necessary nor desirable to fill the bomb completely with liquid below the critical temperature, as the added partial pressures of the trapped air and vapors of the intermediate and transitional fluids can raise the pressure to unsafe heights when the sealed container is heated. Details of use are discussed in a later section.

Fig. 2–2. Original bomb for Freon-critical point drying (Cohen *et al.*, 1968). Rubber tubes at bottom connect to water jacket; armored tube at left is Freon supply tube. Viewing port, pressure gauge, and thermometer are shown. 0.33 X.

Various critical point drying devices have been described by Anderson (1951, 1956, and 1966), Boyde and Wood (1969), Cohen *et al.* (1968), and Fromme *et al.* (1972). The one built by Cohen *et al.* (1968) is shown in Fig. 2–2.

Commercially Available Apparatus

There are currently at least five manufacturers of critical point drying devices; (for addresses see pages 104–105). The details of these devices are presented below.

Bomar (Fig. 2–3a). The Bomar Company makes several models of critical point drying devices. These devices are compact, and have viewing ports, mounted focused lights, and chambers (26 mm in diameter and 25 mm deep). Safety devices include: rupture disks in all models; thermal and pressure switches, which cut off the heat when either temperature or pressure is increased beyond the safe level, in the electronically temperature cycled instruments; and audible warning in the models that are thermally cycled by circulating water. Exit valves open near

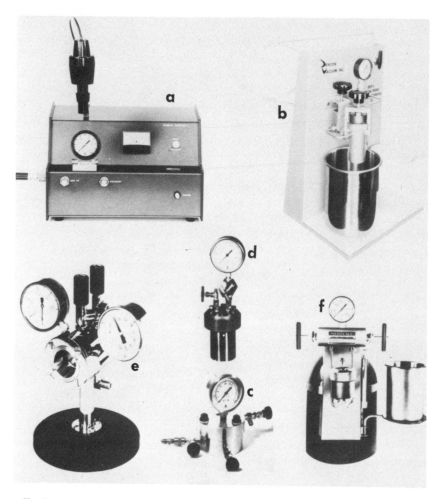

Fig. 2–3. Commercial critical point drying devices. (a) Bomar SPC 900. (b) Denton. (c) Micrographics NCT. (d) Parr. (e) Polaron. (f) Sorvall. See text for descriptions. All photographs supplied by courtesy of the manufacturers.

the bottom of the chamber. Temperature indicators are rapidly responding thermistors. The differences among various models are given below.

Model 900 is for Freon-critical point drying. It has thermoelectric thermal cycling, and needs no water supply. Model 900 EX is for Freon-, CO_2-, or N_2O-critical point drying, and has a greater thermoelectric temperature range. Model 50 is thermally cycled by water circulating through a jacket; otherwise it is similar to Model 900. Model 50 EX is thermally cycled by water jacket; otherwise, it is similar to Model 900 EX. A selection of grid holders and baskets is available. Some are supplied with the instrument.

Denton DCP-1 (*Fig. 2–3b*). The Denton critical point drying device is of rather large capacity (1.125″ in diameter and 2.675″ deep). Safety is provided by a pressure rupture disk. The chamber is sealed with a sturdy yoke which can be hand-tightened. In later models, the exit valve location is warmed by the heating water to prevent its being clogged by carbon dioxide snow when the carbon dioxide-critical point drying method is used. The bomb may be purged from near the bottom by an additional valve and dip tube. Multiple grid holders, stub holders, and stainless steel baskets are available.

Micrographics NCT Line (*Fig. 2–3c*). This is a simple cylindrical box whose lid carries a pressure gauge. Safety is provided by a rupture disk. It is equipped with three valves: the first for inlet, the second for exhaust, and the third for draining intermediate fluid from the bottom. The user furnishes heating and cooling arrangements. Thermal cycling is accomplished by raising and lowering a vessel filled with water having an appropriate temperature around the bomb.

Parr Critical Point Drying Apparatus (*Fig. 2–3d*). This apparatus is sold as separate components which are assembled as a chain leading from the supply tank. Safety is provided by a rupture disk. There is a pressure gauge but no thermometer. The bomb (310 ml capacity) is sealed by a split ring fastened down by four bolts, which are tightened by the wrench provided. Thermal cycling is accomplished by raising and lowering a container of warm or cool water around the bomb. A unique feature of this apparatus is the cooling coil, which can be immersed in a coolant to precool the Freon. A Dewar flask is also provided, presumably to rinse the specimen in the liquid phase of the transitional fluid at a low temperature and atmospheric pressure. The apparatus is suitable for Freon and carbon dioxide. Larger capacity bombs are available and modifications may be made on special order.

Polaron Critical Point Drying Apparatus (*Fig. 2–3e*). This is a water-jacketed device mounted on a pedestal. The chamber is a horizontal cylinder having a viewing port at the end. Safety is provided by a relief valve which blows the gas stream down the hollow pedestal. The device is equipped with pressure gauge and thermometer. Metal blocks are available to take up unused volume and thereby spare expensive transitional fluids. The instrument is completely equipped with grid carrier and containers, some of which are designed to transport delicate specimens into the bomb while they are still immersed in an intermediate fluid.

Sorvall CPD System (Fig. 2–3f). This device is mounted on a panel which clamps directly on the supply cylinder. Safety is provided by a rupture disk. The bomb has a quartz viewing port, a small battery-operated light, and a knurled retaining cap which is screwed on. The internal dimensions are 25 mm in diameter and 25 mm in depth. There is a pressure gauge but no thermometer. The bomb both fills and empties from the bottom. Thermal cycling is accomplished by raising and lowering a container of water at appropriate temperatures around the bomb. No carriers or specimen containers are listed.

Specimen Containers

Containers or carriers for specimens processed through various steps of the critical point drying procedure have received little attention. There are a few descriptions of carriers for microscopic or larger specimens (Anderson, 1956 and 1966; Cohen et al., 1968; Cohen and Garner, 1971; Nemanic, 1972). They are not considered standard components by all manufacturers of critical point drying devices. Generally the investigator is left to his or her own ingenuity in devising containers. Some commercial and homemade ones are shown in Figs. 2–4 and 2–5. Most of the homemade carriers described below and illustrated in Figs. 2–4 and 2–5 may be quickly made in the laboratory without special tools. Nemanic (1972) describes several containers and carriers.

Microscopic specimens obviously require some type of container or carrier, and the handling of larger specimens is facilitated by their use. In the following discussions, specimens will be divided into three size groups which overlap.

Large Specimens. These specimens (e.g., organs, tissue culture sheets, and colonies on agar blocks) are sufficiently large to be easily handled. It is preferable to use a container for specimens even when they can be handled individually. Specimen handling with a pair of forceps at each fluid change is likely to damage the specimen, and materially slows down the procedure. In the later stages of dehydration or in intermediate fluids, many specimens become somewhat brittle. Thus they are easily damaged by manipulation, and the fragments contaminate the surface of the specimen and also the interior of the bomb. The transfer is facilitated by handling a container of specimens rather than by transferring them singly.

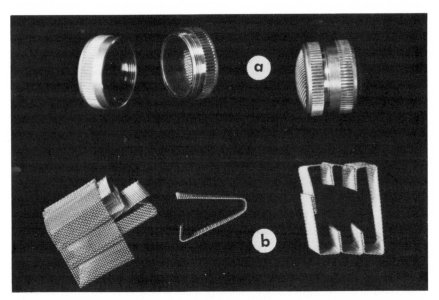

Fig. 2–4. Holders for large specimens. (a) Basket originally sold commercially to hold small watch parts and jewels for cleaning. (b) Homemade shelves and drawers for slender or flat specimens. 1.25 X.

This is particularly important when dehydrated specimens must not be exposed to atmospheric moisture.

When a volatile intermediate fluid (e.g., Freon TF) is used, a container minimizes the danger of evaporation, as the film of liquid carried in its meshes keeps the atmosphere saturated with the intermediate fluid. The amount of intermediate fluid transported by the carrier of small specimens is usually negligible. In the bomb, violent currents are set up during the heating cycle and discharge of the gas phase; unconstrained samples are buffeted and often fragmented.

It is unnecessary, or in some cases even undesirable, that there be free and unrestricted flow through all sides of a container. There should, however, be as free a flow as possible around a specimen in a container. Entrapped air bubbles, as long as they do not completely occlude a portion of the container, do no harm. Restricted flow probably protects specimens against the distortion effects of too rapid fluid exchange (Cohen and Shaykh, 1973). It is necessary that there be no blind pockets in which liquid can remain stagnant and hence contaminate subsequently used fluids.

Representative containers are shown in Fig. 2–4; their construction is, on the whole, self-evident. The "drawers" of 50 mesh stainless steel

Fig. 2–5. Holders and containers for grids, deliciate specimens, and small or microscopic specimens. (a) Commercial holder (Bomar) for grids, retaining screen bent up to show wells for grids. (b) Homemade grid holder, showing base with central post, slotted grid holder, and covering washer. This design (Cohen *et al.*, 1968) allows several holders to be stacked for processing. The design can be improved by deeper wells for the grids. (c) Multipurpose holder made by cutting off the top and bottom Luer hubs of a Swinney filter holder. The ball at the bottom seals the holder when it is lifted for transfer, so that liquid supporting delicate structures does not drain out. Also shown are screens, membrane filters, and Teflon spacer O-ring (see text for further explanation). (d) Simple filter paper cone (for minute specimens), shown open and stapled shut. 1.2 X.

screen (Fig. 2–4b) and their holder are extremely useful for flat specimens (colonies on agar, leaves, and tissue slices) or cylindrical ones
(sections of intestine and stems)—in other words, specimens in which
only one or two dimensions predominate. The cover is bent to form a
wide gap; the specimen is inserted; and the cover is folded back. The
drawer is firmly held with heavy forceps and slid into the carrier.

Useful holders can be made of BEEM capsules cut into cylinders. The
centers of two caps are cut out so that the caps form retaining rings to
hold fine stainless (Nemanic, 1972) or nylon (Cohen and Garner, 1971)
mesh. These are useful for small to minute specimens as long as the mesh
is smaller than the specimen. Two hundred mesh stainless steel screen
seems to be near optimum for most specimens. Similar containers of
larger size may be made by cutting off the ends of polyethylene vials,
cutting out the centers of their caps, and using the caps to hold screens
in place. It is desirable that plastic be tested overnight with the intermediate fluid to be used. Polyethylene is inert to the fluids mentioned
earlier, whereas Tygon and similar flexibilized plastics are adversely affected by these fluids and may leach oily plasticizers into them.

The device shown in Fig. 2–5c is designed to carry very fragile specimens which cannot be lifted out of the liquid without collapse (e.g., root
hairs on roots). It consists of the bottom cup of a Swinney adapter, with
the male Luer fitting cut off, and a stainless steel ball bearing which
falls into the hole in the cup and projects through the bottom. When the
cup is set down, the ball is pushed up, and fluid can then flow through
the bottom hole. Specimens are sandwiched between stainless steel
screens and separated by a Teflon O-ring spacer. Membrane filters may
be used if the pores of the screen are too large. The cap, whose female
Luer fittings have been removed, is screwed on to make a compact capsule. Fig. 2–5c shows the exploded assembly.

Useful and inexpensive metal mesh baskets are commercially available
from watchmaker suppliers (Fig. 2–4a) and other sources. Several types
of baskets are provided for processing of histological specimens. Most of
these are rather large for critical point drying.

Minute Specimens. These specimens (e.g., large protozoa, small in-
vertebrates, large pollen grains, and starfish eggs) are not as large as
the first group, but are sufficiently large to settle out of suspension in a
reasonable amount of time. They do not require centrifugation between
various steps, and are individually visible. These specimens are too small
to be handled individually with ease, but can be handled with pipettes.

Large, or plentiful minute specimens may be carried through processing in the plastic cylinders with the fine mesh covers previously de-

scribed. There is always a tendency for small specimens to be caught in the meshes of the screen, in crevices, and on the inner walls of the cylinders. Therefore sufficiently large numbers of specimens should be available to allow for the attrition.

Alternatively, the specimens may be collected on a membrane filter, of a type not affected by the processing fluids (check manufacturer's specifications for filters), using a Swinney adapter similar to the one described previously and illustrated in Fig. 2–5c, except that the ball seal is removed. The adapter can rest centered over a one-hole rubber stopper in a filtration flask. With mild vacuum, the bottom is sealed against the stopper, and fluid filters through. Care must be taken not to let the filter become dry. The specimens on the bottom filter, while immersed in liquid, may then be covered with a stainless steel screen, and the unit carried through dehydration and critical point drying.

Paerl (1973) reported folding filter disks in half with the trapped material inside. The filters were kept folded through processing by clamps of small pieces of aluminum foil. The writer has found this an elegantly simple and practical procedure.

The 13 mm diameter filter is recommended, because it is a standard size, comes in a very wide variety of filter types and pore sizes, and has the largest range of funnels and adapters available. It is also economical, and an entire disk may be mounted on a specimen stub. Pore diameter should be considerably smaller than the least diameter of the specimens being collected (perhaps no more than one-fifth the smallest dimension) to avoid having elongate cells partly sucked into the pores.

For handling a scant number of minute specimens, Garner and Bryant (1973) have developed an ingenious carrier. Full details are given in their paper; in brief, 3 mm disks are punched out of a membrane filter which is soluble in an intermediate fluid (i.e., acetone). The disks are moistened, placed on a vertical post, and, with a pipette or micromanipulator, the minute specimens placed upon it. Collodion film spread on water in the conventional manner is picked up on the wire loops (5 mm interior diameter), which have been previously prepared. A film is brought down over the disk of membrane filter bearing the specimens, and seals it around the edges.

The assembly is placed under warm, melted 4% agar in a petri dish, held in place horizontally with an applicator stick until the agar partly gels. After complete solidification, a plug of agar containing the collodion membrane and filter surrounding the specimens is cut out. A glass-headed map pin is pushed through the agar to one side of the specimens, thereby simultaneously providing a handle, a weight to keep the plug submerged, and, by the color of the head, a code to identify the specimen. Several capsules may be prepared in the same dish of melted agar. Further treatment is in five-minute changes of water/acetone, 10, 30, 50, 70, 100% (3 changes)/Freon TF(100%)(3 changes)/Freon 13-critical point drying.

Acetone dissolves both the membrane filter and the collodion membrane cover, leaving the objects on agar. It may not be necessary to dissolve the membranes; in which case, ethanol could serve as the dehydration fluid (see discussion of these fluids later).

Larger quantities of specimens may be handled in a very simple funnel of filter paper (ordinary qualitative filter paper is quite satisfactory), as shown in Fig. 2–5d. The suspension is filtered through the paper supported in a funnel; the cone is then removed, folded, and stapled (Fig. 2–5d). Specimens as small as human red corpuscles have been treated in this manner with extremely satisfactory results. After critical point drying, the paper is carefully unfolded and the specimens deposited on the stub or grid for SEM or TEM examination, using any of the procedures described later. We routinely use this simple method developed by Gerald Garner.

Containers such as filter paper packets or agar capsules, which can hold a large amount of liquid in their interstices, have advantages and disadvantages. On the debit side, the total time for fluid exchanges is lengthened, and two or three complete changes of Freon 13 (or 116) must be made in the bomb at no less than five-minute intervals to insure that the T_c of the mixture is in a reasonable range. If CO_2 is used, the conditions are probably more stringent, since the intermediate fluid should be totally removed. On the credit side, replacement of the fluids through the interstices of the filter is gradual, and the danger of osmotic collapse is thus decreased.

Microspecimens. These specimens (e.g., bacteria, fungal spores, small protozoa, pollen grains, and isolated cell organelles) are of microscopic or submicroscopic size (10 μm or smaller). They do not settle out of suspension in a reasonable amount of time, and are individually invisible to the naked eye. Under appropriate treatment, particles in this category may be expected to adhere to the filmed surfaces of grids or other substrates. The first specimens prepared by using the critical point drying method for electron microscopy were bacteria (Anderson, 1951, 1956, and 1966).

Filmed grids are suitable carriers for minute specimens of approximately 20 μm or smaller (exclusive of appendages such as flagella) for either SEM or TEM observation. Grids of 200 mesh are suitable; a smaller mesh is better if the specimens are not too large, but larger meshes increase the danger of film breakage. The substrate can be carbon-Formvar, carbon, or Formvar, with carbon-Formvar being the most stable. Collodion can be used if acetone, amyl acetate, and other collodion solvents are avoided. However, collodion is markedly weakened by ethanol. Filmed grids are excellent for SEM observation as well; the grids are easy to locate on the stub, the bars serve as useful indicators of magnification and perspective distortion, and the film is a smooth, featureless background for the specimens. Loaded grids may be carried on flat, perforated plates, each covered by a cellulose acetate membrane, as described by Anderson (1966), or in wells in holders of the type shown in Fig. 5a and b, covered loosely with a screen or plate. Surprisingly little material is lost with such unsealed carriers. Processing may begin at once, or the loaded grids may be allowed to rest for 3 to 5 min to allow particles to settle and to adhere to the carrier films. In a mixture of particles, some may be adsorbed preferentially on a film. Therefore total or differential counts may be in error (Dubochet and Kellenberger, 1972).

Alternatively, specimens may be concentrated on a membrane filter handled in any of the methods described under minute specimens, or they may be concentrated by putting drops of suspension on slightly dried 2% agar. The water is absorbed, leaving the specimens firmly attached. The agar block may be treated as a large specimen.

THE PROCESSING OF SPECIMENS

Preliminary Treatment of Specimens

In specimen preparation for sectioning for the TEM, tissues are removed quickly to avoid ultrastructural changes due to anoxia, and are quickly cut into very small pieces to allow rapid fixative penetration but with slight regard for surface trauma. These conditions are somewhat eased for critical point drying except for the specimens also destined for TEM study by sectioning (Wickham and Worthen, 1973; Erlandsen et al., 1973), as discussed later. However, the treatment of surfaces becomes more important for SEM study.

Minute and microspecimens should be washed and concentrated by centrifugation or filtration before fixation if: (1) the fluid in which they are suspended can be precipitated by fixatives; (2) their suspension is inconveniently dilute or bulky; or (3) by suspending them in water, they swell to a "normal" size (e.g., pollen grains and spores) (Figs. 2–6 to 2–9).

Large specimens should be handled with care, since every bruise or injury may appear in the electron micrograph. Organs should be cut out with sharp scissors or razor blades and rinsed with water or an ap-

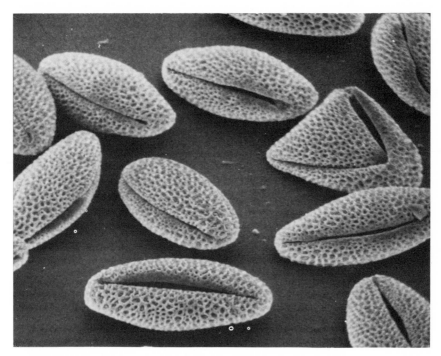

Fig. 2–6. Air-dried pollen of willow (*Salix scouleriana*), fixed with OsO₄ vapor to reduce charging. 1,600 X.

propriate saline. If possible, they should be trimmed to portions not greater in diameter than the support stub and as thin as possible. Soft tissues (e.g., liver, kidney, and intestine) may have preliminary trimming for orientation, and final trimming can be carried out after fixation. Nemanic and Pitelka (1971) used this procedure for mammary glands. If osmium fixation is used, the uniform blackening of the tissue may conceal clues for orientation. In such cases, the preliminary dissection should be definitive if possible. Sketches may be useful to define the section planes.

Delicate specimens which are subject to collapse or damage by repeated removal from liquid and reimmersion must be kept immersed at all times. Such specimens (e.g., roots with long root hairs) should be trimmed under water or saline solution in a bowl sufficiently large for comfortable manipulation of the specimens and sufficiently deep to allow them to be placed in a completely immersed cup (Fig. 2–5c) without being lifted above the surface. The cup may then be transferred through the substituent fluids and into the bomb. The bomb should be thoroughly

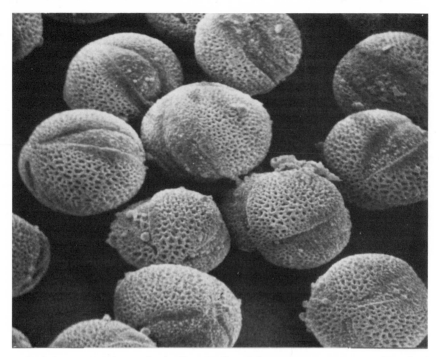

Fig. 2–7. Freon-critical point dried pollen of willow (*Salix scouleriana*). Compare with air-dried pollen in Fig. 2–6. 1,600 X.

purged because of the excessive amount of intermediate fluid carried over.

It is tempting to dehydrate specimens considerably larger than those described above—that is, several centimeters in each direction. Generally this accomplishes little. The examination is not expedited, since it is considerably more difficult to orient oneself to a large specimen than to a small one. Fixation is prolonged and may remain incomplete, and all steps are more lengthy, with increased danger of incomplete removal of the intermediate fluid or incomplete dehydration. The bulkier the specimen, the poorer the electrical conductance (even after coating) and the greater the danger of charging effects will be. Ideally the specimens should have small vertical dimensions and should be attached firmly to the stub by the largest feasible area.

It is impossible to overemphasize that specimens should not be crowded. Flexible or limp fresh organs become fragile dry specimens which shatter or tear at any effort to disentangle or separate them from one another.

Fig. 2–8   Fig. 2–9

Figs. 2–8 and 2–9. Air-dried (Fig. 2–8) and critical point dried (Fig. 2–9) spores of *Badhamia utricularis* (Myxomycete). In both cases, the spores are shown clinging to branches of the capillitium. Note in Fig. 2–9 that spore clusters, typical of this species and important for diagnosis, did not disperse during critical point drying. 2,500 X.

Specimen Cleaning

Fresh tissue under the high power of the dissecting microscope shows vivid colors. Plant cells are turgid and often shiny. Animal tissues show the delicate translucency of membranes and red capillaries in them. In sections of viscera, the tubules, layers, villi, and other structures are beautifully and intricately arranged. It is therefore surprising to see these tissues in the SEM appear as if they were covered with limp, opaque plastic sheeting or obscured by gluey strands or clots. In the TEM, the bacteria (which were visible as distinct circles or rods under the light microscope) are barely discernible in a muddy background. The reason for this is simple. All matter is somewhat opaque to electrons. Therefore gelatinous capsules of bacteria, invisible in the light microscope, become a murky halo in the TEM. The mucus, plasma, gum, or wax covering plant or animal tissues, although transparent to light waves, forms the opaque surface seen in the SEM.

In the course of specimen preparation, the dehydration and intermediate fluids may dissolve some surface components, particularly waxes. On the other hand, they fix and toughen proteins, polysaccharides, and muco-polysaccharides. Although there is no single universal method for cleaning the surfaces, a thorough rinsing of surfaces (which are normally moist) with water or an appropriate saline solution should always be carried out. The vigor of the rinsing depends upon the delicacy of the surface. Ciliated epithelium, for instance, should receive gentler care than intestinal mucosa. The lumens of cavities (intestine, blood vessels, and heart) should be flushed out. Suspensions of cells in plasma or organs whose surfaces are moistened with completely liquid plasma are sufficiently cleaned after rinsing.

Normally surface-dry specimens (leaves, petals, and stems) should be cleaned with puffs of air or gas from a duster can, the forcefulness of such cleaning being dependent upon the hardiness and dirtiness of the specimen. Additional methods that have been used are presented below.

(1) Specimens have been treated before or after fixation with enzymes. Examples are: papain and other enzymes for various tissues (Boyde, 1972; Boyde and Wood, 1969); chymotrypsin for intestinal mucosa (Kanazawa et al., 1972); trypsin for isolation and cleaning of parietal cells of stomach mucosa (Lee, 1972); diastase "and other enzymes," not specified, for cleaning organelles isolated by mechanical means (Sandborn and Makita, 1972); trypsin for chromosomes (Wolfe and Martin, 1968); and pronase and elastase for removing neuronal sheaths (Lewis, 1971).

(2) Specimens have been treated with other reagents. Examples are: a prolonged treatment (after glutaraldehyde fixation) with sucrose for gastrointestinal mucosa (Landboe-Christiansen and Parapat, 1972); isolating cells of hydra by omitting Ca and Mg from medium but adding sucrose (Westfall and Enos, 1972); treatment with dilute ethylene diamine tetraacetate (EDTA) for separating nerve cell processes (Lewis, 1971); prolonged maceration with glycerol and dilute ethanol to remove milk from lactating mammary glands (Nemanic and Pitelka, 1971); the use of N-acetyl cysteine as a mucolytic agent (Williams et al., 1973); and the use of 5% HCl to dissolve the gelatinous layer around sensory cilia and cupulae of the semicircular canals (Lim and Lane, 1969).

(3) Specimens have been subjected to mechanical treatment. Examples are: a jet of compressed air to remove mucus from sensory papillae of tongue (Shimamura and Tokunaga, 1970); and sonication of cornea and other tissues after a brief fixation with osmium tetroxide (Kuwabara, 1969 and 1970).

Fixation

Certain specimens can be dehydrated and critical point dried without prior fixation. Examples are: pollen grains (Garner and Bryant, 1973), moss leaves (Cohen and Garner, 1971), and isolated organelles (Arget-singer, 1964; Pihl and Bahr, 1970; Wolfe, 1965). Probably most tissues and isolated cells require fixation, which provides more stabilization than is provided by the dehydration fluids. The goals of fixation are: to toughen the specimen to withstand the physical stresses of subsequent treatment; to preserve natural morphology; to preserve internal cellular anatomy; and to arrest, without agonal changes, physical activities such as ciliary and flagella beating and the position of legs in arthropods.

The goals mentioned above are relevant for most specimens as shown strikingly for such tough specimens as pine needles. The pine needle of Fig. 2–12 was exposed to osmium tetroxide fumes, which penetrated only through the cut end and not through the waxy cuticle. On processing through water/ethanol/Freon TF/Freon 13-critical point drying, the fixed portion retained original dimensions, whereas the unfixed part showed some collapse. The undistorted segment corresponded to the part of the needle blackened by the fixative, and the shrunken segment corresponded to the part which was still green.

Fixatives themselves may produce the very changes they are supposed to avoid; these changes include agonal movements and osmotic distortion. These problems will be discussed in the following sections. Although dehydration by fluid substitution is itself a fixation, when "fixation" is mentioned in this chapter, it refers to the prior treatment and not to the effects of dehydration reagents.

The fixatives extensively used for critical point drying are osmium tetroxide, osmium tetroxide in combination with mercuric bichloride (Parducz's fixative), formaldehyde, and glutaraldehyde. The selection of fixative depends largely on previous TEM experience (see Hayat, 1970, for a detailed discussion) or arbitrary choice. However, the need for information on fixation for scanning electron microscopy is recognized, and there has been active research in finding optimal conditions and developing a rationale for fixation (Arnold et al., 1971; Boyde and Vesely, 1972; Cohen and Shaykh, 1973; Falk et al., 1971; Nowell et al., 1972; Panessa and Gennaro, 1972a).

Aldehydes. Of the aldehydes, only glutaraldehyde and formaldehyde have been used to any extent as fixatives for critical point drying. They are best suited for relatively large specimens of animal tissues, injection or perfusion of organs (where later enzymatic digestion is contemplated),

Fig. 2–10   Fig. 2–11   Fig. 2–12

Figs. 2–10 and 2–11. Air dried and critical point dried stamen hairs of *Tradescantia*. The tamen hair of *Tradescantia* is a chain of individual thin-walled cells. Air-dried cells collapsed, while cells fixed with OsO_4 vapor and critical point dried by the Freon method retain their shape. 250 X.

Fig. 2–12. Partially fixed needle of pine (*Pinus ponderosa*). Vapors of OsO_4 entered from the cut end only, and did not penetrate through the cuticle. The fixed portion retained its dimensions during dehydration; the other portion collapsed. 42 X.

and double fixation with osmium tetroxide for specimens intended for sectioning as well as critical point drying (Wickham and Worthen, 1973).

While the aldehydes penetrate between cells more rapidly than does osmium tetroxide, they enter into the cells more slowly. The rate of fixation is also slow. An advantage of slow fixation is the decreased danger of overfixation. Aldehyde solutions are reasonably stable and therefore suitable for field use. Specimens may be processed days or weeks later. The disadvantage is long duration (hours or days) of fixation required for some specimens. Glutaraldehyde is considered slower than formaldehyde.

In a study of speed of fixations, Wohfarth-Bottermann and Komnick (1966) found protoplasmic streaming or other visible activity continuing seconds or minutes after glutaraldehyde addition. In our laboratory, nematodes continue to move for at least 1/2 hr in 3% glutaraldehyde, and small arthropods (mites and freshwater crustacea) continue movement for minutes. Glutaraldehyde is therefore not suitable for immediate arrest of motion and maintenance of a "natural" appearance.

Glutaraldehyde and formaldehyde in useful concentrations retain considerable tonicity, compared with osmium tetroxide of equivalent fixation ability. Therefore their solutions alone or with buffer may be hypertonic. Although the shrinkage caused by these solutions is usually little noticed in TEM sections, it can be disastrous for surface aspects of many specimens. The problem is compounded because of the ability of glutaraldehyde to preserve at least some of the selective permeability of the cell membrane (Jard et al., 1966). Cells thus remain sensitive to the osmotic variations which occur in the course of fluid substitution (Cohen and Shaykh, 1973). Nevertheless, acceptable to excellent preparations of some higher plant (Panessa and Gennaro, 1972a) and many animal tissues have been obtained by using aldehydes.

Osmium Tetroxide. Osmium tetroxide was probably the first fixative used to prepare multicellular specimens by critical point drying (Overman, 1963). Its drawbacks are: great expense, sensitivity to organic contaminants, irritating vapor, slow penetration between cells of tissues, and the blackening of specimens, which makes orientation difficult. Its advantages are: usefulness in a very low concentration, little osmotic effect in low concentrations, versatility (it may be used in vapor form as well as in liquid form), extremely rapid fixation of cells, and concomitant immediate arrest of cellular activity (Wohlfarth-Bottermann and Komnick, 1966), increase in cell membrane permeability (thereby making cells less sensitive to the tonicity of the fixative and subsequent fluids), and reduction of charging artifacts in SEM specimens (Pfefferkorn, 1970).

The ability of osmium tetroxide to preserve delicate features of cell morphology was noticed as early as 1927 by Strangeways and Canti in their thorough study of the effect of fixatives on tissue cultures. Of the fixatives (chromic acid, potassium bichromate, mercuric chloride, picric acid, alcohol, acetone, formalin, and acetic acid) used in various strengths and combinations, osmium tetroxide was the only one producing no detectable shrinkage. It is the writer's opinion that osmium tetroxide alone or sequentially with other fixatives is probably the most generally useful fixative for critical point drying.

Physical Fixation (*Quick-Freezing*). Although osmium tetroxide stops cellular activity almost instantaneously, it must reach the cells to do so. Arthropods, even delicate and microscopic mites, aphids, and water fleas, are highly impervious to the liquid or fumes, and may require as long as 1/2 hr to be immobilized. When they are finally killed, the appendages are usually tightly flexed or contracted. In our laboratory, these animals are arrested in a normal condition by plunging them, while they are normally active, into liquid N_2 or Freon 22 cooled to liquid N_2 temperature. The specimens can be dehydrated gradually by transferring them to ethanol cooled to $-20°C$ (standard deep-freeze temperature) and maintaining them in a freezer while the ethanol gradually dissolves and replaces the frozen water. Subsequent treatment is conventional. If the organisms are allowed to thaw without the alcohol treatment, they often show irregular movement and final tight flexure of the appendages.

Osmotic Effects

The fixation sets the upper limit to the quality of the final preparation, and is the source of many artifacts, presumably of an osmotic nature. There is a major difference between plant and animal tissues in the dimensional effects of fixation and subsequent treatment (Cohen and Garner, 1971; Cohen and Shaykh, 1973). Most animal cells are surrounded by a cell membrane, which may either be naked or have only an external mucoid coat. The cell membrane is flexible and elastic. Therefore it can stretch considerably before breaking, and contract appreciably without wrinkling. Plant cells usually have a cell wall external to the cell membrane. Except when the cell wall is highly lignified or very thick, as in woody or cutinized tissues, it is very flexible but almost inelastic. Therefore, while it can constrain the cell contents against swelling, the cell wall can wrinkle and collapse if the contents shrink. The distortion is not so evident in a bulky tissue (e.g., parenchyma), where each cell supports and is supported by its neighbors; however, it is evident in single cells and exposed cells of the outer layers of tissues.

The larger the individual cells, the more sensitive they are to distortion during fixation and fluid exchange, since volume stresses increase, as discussed earlier. Also, osmotic equilibrium is reached more slowly with large cells. For this reason, very small or narrow cells (bacteria, yeasts, and fine fungal hyphae) may, with little or no fixation, tolerate considerably more osmotic shock in hypertonicity and abrupt changes between solution concentrations than do larger cells.

If the fixative is too strong and concomitantly hyperosmotic, it produces shrinkage and distortion. If it is weak, in order to avoid such distortion, subsequent treatments can distort the inadequately hardened specimens. Thus light fixation preserves initial shape but allows it to collapse later. On the other hand, strong fixation causes shrinkage and distortion but allows faithful preservation of the artifacts.

The solution to this problem appears to lie in sequential fixation, either with different fixatives or with increasing concentrations of the same fixative. In a study of normal and malaria-infected erythrocytes and bacteria, Arnold et al. (1971) tried several combinations of buffer, glutaraldehyde, and osmium tetroxide. Their specimens were air-dried from ethanol or propylene oxide (used as intermediate between the ethanol and Epon embedment for sectioning). They found that any single fixation (e.g., osmium tetroxide in any concentration, 0.1% glutaraldehyde, or 2% glutaraldehyde) caused distortion. However, fixation with 0.1% glutaraldehyde/2% glutaraldehyde/osmium tetroxide (concentration not given)/ethanol/propylene oxide/air-drying provided the best preservation of shape. Bessis and Weed (1972) also used glutaraldehyde/osmium tetroxide/ethanol/propylene oxide/air-drying for red cells, and showed striking correlations between the physiological state of the cell and its final shape. Propylene oxide apparently acts as a fixative.

Vesley and Boyde (1973) used the following double fixation for tissue cultures of fibroblasts and tumor cells: glutaraldehyde 1% in 0.15 M Na cacodylate, pH 7.4 "with added Ca and Mg ions" (37°C, 1 hour)/glutaraldehyde 3% in 0.15 M Na cacodylate, pH 7.4 (6°C, 2 to 6 days)/H_2O-ethanol/Freon TF/Co_2-critical point dried. Starting with an even more dilute fixative, Stewart et al. (1973) perfused canine jugular veins with Tyrode's solution, then fixed them successively in 0.1%, 1%, and 2.5% glutaraldehyde in Tyrode's solution, followed by dehydration and critical point drying with CO_2.

Enlander et al. (1973), after trials with higher concentrations of fixative, adopted 1% glutaraldehyde in cacodylate buffer, followed by postfixation in 1% OsO_4/ethanol/Freon 13-critical point drying for the demonstration of fine surface detail in Herpes virus infected tissue culture cells.

Panessa and Gennaro (1972a) used prolonged fixation (as much as two months) with 3 or 6% glutaraldehyde to harden plant tissues for subsequent treatment; the lower concentration was used when the higher one produced plasmolysis. They also employed uranyl salts subsequently, apparently for both fixation and reduction of charging artifacts in the SEM. Hayat and Giaquinta (1970) employed osmium tetroxide vapor/glutaraldehyde/osmium tetroxide for TEM studies.

Sequential fixation either with the same fixative or with different ones needs further exploration. If each step increases the ability of the cell to withstand greater osmotic differences, one might start with a delicate osmium fixation (perhaps by vapor), followed either by aldehyde or by higher concentrations of osmium tetroxide with the assurance that cells will retain normal shape.

Maser *et al.* (1967) have rendered a valuable service by presenting the osmolalities of fixing agents and buffers as curves of osmolality plotted against concentrations. *The Merck Index* (8th ed., 1968, pp. 1281–98) tabulates the osmotic pressure of many chemicals in useful concentrations.

Dehydration and Intermediate Fluids

Although possibly other water-miscible liquids may serve for dehydration (e.g., tertiary butanol and isopropanol), ethanol and acetone are the major dehydration liquids. They may be used as both dehydration and intermediate liquids for Freon-critical point drying, but the advantage of Freon TF as an intermediate (Cohen *et al.*, 1968) generally limits these liquids to the dehydration steps in the Freon method.

It is possible to go directly from ethanol to carbon dioxide, contrary to previous belief, with the consequent elimination of amyl acetate as an intermediate fluid (DeBault, 1973; Enlander *et al.*, 1973; Lewis and Nemanic, 1973; Mueller *et al.*, 1973).

Most workers use alcohol as a dehydration agent, although a few (e.g., Smith and Finke, 1972) prefer acetone. The advantages of acetone are: its use requires no special licensing, it is not as hygroscopic as is ethanol, and it mixes easily with Freon TF. However, we have found that the cell surface may show fine wrinkles not visible with ethanol treatment, and some spores and pollen grains show drastic wall alterations. These may be due to the great solubility of surface waxes and lipids in acetone, but not all surface alterations can be ascribed to the removal of lipid material. Lewis and Nemanic (1973) found severe surface damage of tadpole epidermis when acetone was used for dehydration, but none when ethanol was so used. Daniel (1973) studied ovaries, fat bodies, and bacteroids of the cockroach, *Periplaneta*, after dehydration through acetone with direct transfer to CO_2 for critical point drying. The surfaces of tissue cells in their micrographs appear somewhat wrinkled, although the bacteroids appear nearly "normal"—that is, with only slight wrinkling of the surface.

Acetone cannot be used if the support is collodion, and it weakens Formvar membranes. It is now rarely used for critical point drying in our laboratory.

Ethylene glycol was used by Pihl and Bahr (1970a) to provide an "inert" dehydration, and cellosolve (2-ethoxyethanol) was used as the intermediate fluid for carbon dioxide-critical point drying. This promising technique will be discussed in the following section.

In a study of dehydration-induced artifacts, Cohen and Shaykh (1973) experimented with *Spirogyra* cells (selected as sensitive test objects because of their tendency to collapse during dehydration) fixed with 0.5% osmium tetroxide followed by 1% osmium tetroxide.

In some cases, the specimens were treated with periodic acid to bleach the cytoplasm. These cells showed the least tendency to collapse, but showed the most drastic changes in the cytoplasm. It was found that the higher concentrations of alcohol must be closely spaced, as must the initial ethanol-Freon TF mixtures. To minimize shrinkage and distortion (after fixation in 0.5% OsO_4 and decolorization in periodic acid), it was found necessary to use steps no greater than: ethanol, 15, 30, 40, 40, 60, 70, 80, 90, 95, 100%/Freon TF, 6, 12, 25, 37, 50, 60, 70, 80, 85, 90, 95, 100%. The large step from the intermediate fluid (Freon TF) to the transitional fluid (Freon 13) apparently had no adverse effect. The writer observed *Spirogyra* through the viewing port of the apparatus shown in Fig. 2–2; after the Freon TF was drained, Freon 13 was quickly admitted. No changes were visible at 35× magnification.

A continuous exchange is the alternative to the tedious stepwise treatment with 12 to 30 changes before the specimen is transferred to the bomb. The apparatus shown in Fig. 2–13 was designed for this purpose. The specimen holders (the grid holder of Fig. 2–5b and the container of Fig. 2–4a are shown) rest on a screen about 5 mm above the bottom. After all valves have been closed, the cylinder is filled with water up to or slightly above the level of three-way valve A, the specimens are introduced, and 95% ethanol is carefully layered over them. A gradient is established by gently moving a glass rod up and down through alcohol and the upper water level. Valve B is now slightly opened to allow the liquid to flow out, and the specimens are thus dehydrated through a continuous gradient of water/ethanol. The liquid may be allowed to drop below the level of valve A, and 100% ethanol is allowed to enter slowly from the separatory funnel above valve A to form a distinct layer.

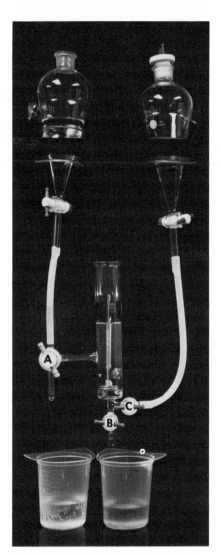

Fig. 2–13. Apparatus for continuous exchange of dehydration intermediate fluids. See text for explanation. (Courtesy of IITRI, SEM/1972). 0.2 X.

The top of the cylinder is kept loosely covered to exclude moisture. It is helpful to add a drop of alcoholic 1% eosin to the alcohol in the separatory funnel; its fluorescence defines the 100% ethanol layer. With bulky tissues, two changes of 100% ethanol should be made. The dehydration is ended by closing valve *B* and leaving 100% ethanol completely covering the specimens. The separatory funnel above valve *C* is filled with Freon TF, and the valve of the separatory funnel and valve *C* are slightly opened to allow Freon TF to rise from the bottom of the cylinder. Valve *A* is opened to allow the fluid to drain. Although a sharp boundary is visible between Freon TF and the alcohol, a gradient does exist. This gradient is steep; therefore for very delicate specimens (e.g. algal filaments), the exchange is accomplished with successive small amounts of Freon TF: ethanol (25, 50, 75, 100% [twice]) allowed to pass through rather quickly—that is, 25 min for all exchanges depending on specimen volume and density.

The rate of fluid substitution depends on the volume and density of the specimen. Approximately 15 min each for water/ethanol and ethanol/Freon TF is sufficient for minute or microscopic specimens; larger ones, such as the intestine (Fig. 2–18), require ~30 min. No apparent damage is done if specimens are left overnight in 95% or higher fluids. We have stored tissues for weeks in Freon TF without apparent damage.

The device is applicable to any other combination of dehydration and intermediate fluids, due regard being paid to the order of relative densities in filling the separatory funnels. The white tubing shown in Fig. 2–13 is Teflon; polyethylene would probably do as well. It should be mentioned again that a plastic should be allowed to soak overnight in the dehydration fluid, the intermediate fluid, and a mixture of the two; only if it is unchanged may the plastic be considered safe for use.

Choice of a Transitional Fluid

Since nitrous oxide does not completely replace the water in specimens and is apparently not completely miscible with water (Boyde, 1972; Nemanic, 1972), it will not be considered. The general properties of the other transitional fluids (carbon dioxide and the fluorocarbons) have already been discussed and presented in Table 2–1. As to their effects on the specimen, Boyde (1972) and Nemanic (1972) claimed no perceptible difference between drying from carbon dioxide or Freon 13. In our laboratory, specimens dried from Freon 13 seem to show less shrinkage on the whole and less fine surface change than those dried from carbon dioxide; however, these impressions may not survive careful and controlled study. The major differences between Freon- and carbon dioxide-critical point drying procedures lie in the tolerance for intermediate fluid and the pressures necessary for critical point drying. The fact that all traces of Freon TF do not have to be removed makes the use of Freons considerably faster than the conventional carbon dioxide method, which requires the practically complete removal of amyl acetate. Vesley and Boyde (1973) have successfully used Freon TF as an intermediate fluid for the carbon dioxide method in a study of tumor cell tissue cultures. If their method is generally applicable, and if the Freon TF does not have to be purged, the carbon dioxide method would then approach the convenience and speed of the Freon method. The lower P_c of Freon 13 (38.2 atm) and Freon 116 (29.4 atm) compared to carbon dioxide (72.9 atm) means components of lower pressure ratings may be used. The use of a viewing port is no longer the exclusive convenience of the Freon critical point drying, as commercial devices with ports suitable for use with carbon dioxide are available.

The fluids used in the Freon processing are not toxic or unpleasant to work with; the amyl acetate of the carbon dioxide method essentially requires a hood. A further advantage of the Freons is the ease with which they are vented. Carbon dioxide must be vented quite slowly (or the outlet port warmed), as it tends to freeze by adiabatic cooling on expansion and to block the outlet. While a rather extensive and prolonged flushing must be carried out to remove amyl acetate, little or no flushing is needed in the Freon TF intermediate/Freon transitional procedure. Therefore, for chambers of 25 ml capacity or less and for small specimens, the cost of the Freon method is comparable with that of the carbon dioxide method. For larger bombs and large specimens, carbon dioxide is more economical. It is also more easily obtainable.

Smith and Finke (1972) advocated the use of Freon 116 instead of Freon 13 because of its lower P_c (29.4 atm) and T_c (19.7°C). In order to effect a change of fluids at atmospheric pressure, the Freon 116 was condensed to liquid by cooling with liquid N_2. The writer believes that such a practice will have the danger of condensing water and hence damaging the specimen. Freon 116 would have to come into a bomb cooled below 19.7°C if it is to enter as a liquid. It may not be convenient to cool a bomb to 10°C to allow rapid filling. Freon 13 has a critical temperature of 28.9, and therefore may be filled at room temperature. However, Freon 13 is approximately twice as expensive as Freon 116.

The Procedure

Initially it is worthwhile to make a few practice runs in order to become familiar with the routine of opening and closing the valves and interpreting gauge readings. The latter is particularly important if the bomb does not have a viewing port. Irrespective of the type of transitional fluid used, the bomb must be dry and at least ten degrees cooler than the critical point. The cooler it is, the more rapidly and smoothly it fills after sealing. In humid climates, precautions must be taken to avoid condensation of moisture in the bomb. It should be kept closed until the specimen is transferred.

The specimens should be ready in a closed wide-mouth jar, immersed in the intermediate fluid. They should be at approximately the same temperature as the bomb, if possible. Care should be taken that the lining of the jar cap does not come in contact with the liquid if it contains waxes or plastics for which the intermediate fluid is a solvent. Also, a desiccator or at least a dry wide-mouth jar with a tight cover should be ready to receive the specimens.

(1) Inspect gaskets and sealing surfaces to make certain that they are clean and free from cracks or other defects which could prevent tight sealing. All valves must be closed. Quickly introduce the specimens, check for any projecting portions which could be caught in the seals, and tightly close the bomb. Open the tank valve and inlet valve of the bomb and allow the fluid to enter gradually. When the pressure is near or above the P_c, the bomb is filled with liquid, provided the temperature is below the T_c.

Some workers turn the tank upside down to get the liquid transitional fluid. This is unnecessary and often undesirable. If the fluid enters as a gas and the bomb is below T_c and the pressure is above P_c, it will condense at once. If it enters as a liquid and the pressure is too low or the temperature is above T_c, it will evaporate. Any foreign material in the tank will be carried over from the inverted tank. Some lots of Freon are contaminated with an oil; oil-free Freon can be obtained at no extra cost on request.

(2) For carbon dioxide, open the exit port slightly and let the gas escape at a rate which balances its entrance and keeps the pressure constant. The bomb may be considered purged when the odor of amyl acetate is faint; it rarely disappears completely. For large specimens, it may be useful to hold the bomb sealed for 15 to 30 min to allow equilibration and then to renew the flow.

(3) For Freon and large specimens or small ones in filter paper, agar, or other fluid-retaining carrier, close the entrance valve and slightly open the exit valve. The pressure will immediately drop but will remain at a constant value, somewhere close to 450 lb psi for Freon 13 at 20°C. If the pressure begins to drop below this level, immediately admit more Freon and close the exit valve. If the bomb has a viewing port, the liquid level can be easily monitored. The pressure will remain constant as long as the liquid is in the bomb and the temperature is constant. Small specimens which have been carried through Freon TF, and which retain very little fluid, need no flushing.

(4) Raise the temperature. In some commercial devices, this occurs at a fixed and suitable rate; otherwise it should be monitored. It should take at least 5 min for the temperature to rise from ~10°C below the T_c to ~10° above it. A slower rate does no harm, whereas too rapid a temperature rise causes violent currents, boiling, and a cycle of recurrent evaporation and condensation until the contents are at equilibrium.

As the temperature rises, if the bomb has a viewing port, it can be seen that the liquid phase expands until either it fills the bomb or, if there was sufficient initial gas volume, the increasingly faint meniscus between liquid and gas vanishes. It is difficult to be certain whether the chamber contains a very dense gas or a very mobile liquid at the critical temperature. It may be necessary to open the exit valve slightly to keep the pressure within safe limits but above the P_c. This would be 900 to 1,000 lb psi for the Freons, and ~1,400 psi for carbon dioxide.

(5) Keep the temperature at least 10° above T_c (higher if there is reason to believe that 5 to 10% of Freon TF is in the bomb) and open the exit valve. Approximately 100 psi/min decline is a desirable rate. If the chamber has a viewing port, and there is a considerable amount of intermediate fluid carried over, the atmosphere inside the bomb will appear hazy. This is a sign that phase separation is about to take place. If possible, raise the temperature ~5°C and decrease the rate of decompression. Very rarely will there be any condensation. A possible explanation is that with the gas phase saturated with the intermediate fluid, rapid decompression cools and supersaturates the vapor. Slower decompression allows temperature equilibrium to be maintained.

Although most specimens can be decompressed at 100 psi per minute without distortion, particular care must be taken with small arthropods whose poorly permeable chitinous exoskeleton does not permit immediate equilibrium. If the construction of the critical point drying device allows it, lead a tube from the exit valve and immerse free end in a cylinder of water. The bubbling rate serves as a very convenient monitor. However, by either an interposed trap or watchful care, eliminate any possibility of sucking water back into the bomb, for instance, by allowing it to cool with the cover sealed on.

Damage to a specimen is not incurred because of its being resubmerged in the transitional liquid but because of its drying by evaporation. If there should be a pool of liquid in the chamber, it should be carefully flushed out. With pressure still elevated, the bomb is brought below T_c, refilled, and the cycle repeated. Many specimens tolerate this emergency treatment rather well.

(6) Remove the specimens and transfer them rapidly to the desiccator or a jar. They are very fragile and brittle, and must be handled with

extreme care. If possible, tip them out of the container instead of pulling them out with forceps.

As the supply tank is used up, the pressure drops and the flow becomes sluggish. There is little economy in trying to get the last of inexpensive carbon dioxide out of the cylinder, but the Freons are charged to such a low pressure that under ordinary circumstances as much as half the supply may remain in the reservoir when the pressure has dropped below P_c. The Freon supply cylinder can be immersed in a bucket of warm water or cautiously warmed with a heat gun kept in constant motion. If practicable, the closed bomb may be cooled below the usual starting temperature to create sufficient pressure differential.

Selected Examples

Arthropods and Other Larger Organisms. Insects and other arthropods larger than the minute class generally need no particular treatment. Conventional entomological methods can be used to arrange legs and wings in natural position before the insects are dried. Critical point drying is not necessary except for specimens with soft, plump abdomens which shrivel when air-dried. Smaller specimens can be arrested by plunging into liquid nitrogen or Freon 22 with further treatment as described previously.

Ciliated and Flagellated Organisms. Parducz (1967) developed a fixative designed to immediately arrest ciliary movement in ciliates. Cells are concentrated by mild centrifugation and, if possible, washed twice by centrifugation with distilled water. Four volumes of the fixative are rapidly added to one volume of concentrated cells, and the duration of fixation must last at least 15 min to harden the cilia sufficiently. Parducz's fixative is prepared as follows:

Aqueous 2% osmium tetroxide	6 parts by volume
Aqueous saturated $HgCl_2$	1 part by volume

The above fixative has produced excellent results when followed by freeze-drying or critical point drying of protozoa (Small and Marszalek, 1969) and the organs of higher organisms (Marszalek and Small, 1969). The front cover of *Science* for March 7, 1969, has the well-known picture of the ciliate *Didinium* ingesting *Paramecium*. Another well-known example is shown in the cover illustration of *Science* for February 21, 1969, illustrating the arrested metachronal waves of the cilia of *Opalina*. Horridge and Tamm (1969) fixed swimming *Opalina* by pipetting 2%

osmium tetroxide rapidly on them. The procedure was: osmium tetroxide (2%, 0°C, 10 to 15 min)/"graded ethanol series"/amyl acetate/carbon dioxide-critical point drying. Earlier, Satir (1963) had successfully used 1% osmium tetroxide for fixing the metachronal waves of clam gill cilia.

Thus equally good results seem to be possible with osmium tetroxide alone as with Parducz's fixative. Apparently no critical studies have been made on the change in dimensions of protozoa or other cells fixed with Parducz's fixative, and in most cases surface irregularities are obscured by the beautifully preserved and often numerous cilia. In the absence of critical comparative studies, 1 to 4% osmium tetroxide alone (i.e., without mercuric chloride) should be the fixative of choice.

The writer has used osmium tetroxide vapor fixation and a dilute fixative (0.1% osmium tetroxide in distilled water) to preserve myxomycete swarm cells in a normal configuration. Except for their immotility, they were almost indistinguishable from living cells when examined as wet mounts under phase contrast. On the other hand, glutaraldehyde or formaldehyde in any significant concentration distorts the shape.

It is essential that exposure to the liquid fixative or concentrated vapors be a rapid one. Contaminant traces of fixative or other toxic materials can cause rounding up, contraction, or other distorting changes before fixation. The source of such a contamination is not merely an open container of fixative, but forceps and other instruments, slides, and even the copper of grids on which suspensions of live organisms are placed.

The following method for rapid osmium tetroxide vapor fixation of motile cells is recommended. Prepare a fixing chamber by piercing the rims of the cover and bottom of a small plastic petri dish (60 mm diameter) with a hot nail so that when assembled the two holes coincide. Place six drops of 4% osmium tetroxide on the bottom (away from the hole) of the dish and turn the cover to close the opening. Meanwhile, have filmed grids (carbon, Formvar, or collodion) prepared and the cell suspension ready. Grids, specimen suspension, and instruments should be as far from the vapor chamber as one can comfortably reach, and of course with no draft blowing from the chamber to these objects. Transfer a small drop of the suspension to the grid (which should be wetted by the drop) so that the drop forms a slightly convex shape instead of a hemisphere (use drawn-out Pasteur pipette or bit of filter paper to remove excess suspension). Rotate the chamber lid to make the holes coincide, and immediately introduce the grid. It can be left in the chamber while other grids are prepared. Rinse and dry the forceps or use a fresh pair before the next grid is introduced. Prepare each grid immediately before use, allowing enough time for organisms disturbed by pipetting to resume normal activity. Do not let grids stand with un-fixed suspension; otherwise cells will be damaged by fumes, or traces of copper from the grid itself.

The above method can be modified to accept coverslips or slides by using large petri dishes with sectors cut out of the rim and with a small open dish inside the large one to hold the osmium tetroxide. The fixative should be renewed after 30 min. This method minimizes both the contamination of the surrounding atmosphere and the dilution of the fixative vapors which occurs when a cover or stopper is removed with unavoidable pumping or fanning action.

Microorganisms and Other Small Cells. In his original work, Anderson described the critical point drying of bacteria (*Escherichia coli* with page attached, *Spirillum sp.*), red corpuscles, and trichocysts of *Paramecium*. Drops of suspension were placed on filmed grids, which were then exposed "for a few minutes" to the vapors of 2% osmium tetroxide. Covering the grids with a film of cellulose acetate, as described by Anderson (1956 and 1966), is not necessary if the grids are held in covered holders such as illustrated in Fig. 2–5a and b. Microorganisms adhere well to the grid film.

Bacteria can be fixed, before the suspension is placed on the grids, filters, or other carriers, with osmium tetroxide or dilute (2 to 5%) formalin. If they are to be examined on the same substrate on which they are collected (i.e., filmed grids), it may be preferable to fix them on the substrate with osmium tetroxide vapors. Figure 2–14 is a stereopair of TEM graphs of *Pseudomonas fluorescens* prepared by this method. Mackey and Morris (1972), using the Ryter-Kellenberger (OsO₄) method, have shown spores suspended by fine strands within bacterial cells.

Bystricky *et al.* (1972) studied the complex external morphology of a peculiar soil bacterium with both the SEM and the TEM. For SEM, colonies on small blocks of agar were exposed to OsO₄ vapor for 24 hr, and then passed through ethanol and dried by the Freon-critical point method. For TEM studies, Formvar-coated grids were touched to the unfixed colonies, dehydrated without prior fixation, and dried as described. Grids were chrome-shadowed. OsO₄ fixation was used for the SEM preparations to minimize charging artifacts (Pfefferkorn *et al.*, 1972), and omitted for the TEM preparations to retain transparency of the cells.

Organelles. Organelles are obtained from cells by using a variety of methods: mechanical chopping or grinding, enzymatic dissolution of a

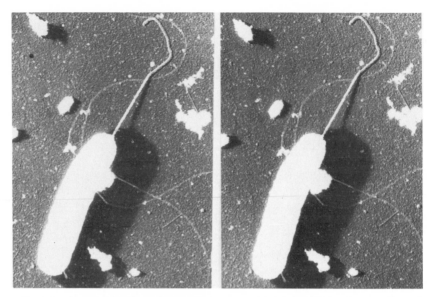

Fig. 2–14. A stereopair of *Pseudomonas fluorescens*, critical point dried with Freon. Transmission electron micrograph (courtesy of R. G. E. Steever). 23,000 X.

surrounding matrix, or the use of surfactants to disperse the plasma membrane and the cytoplasm. Despite a vast literature on the subject (Deter, 1973), surprisingly little has been published on the elucidation of organelle structure with the aid of critical point drying. Chromosomes are an exception to this generalization, for there has been an active investigation of the chromosome structure using critical point drying.

Argetsinger (1964) isolated basal bodies of cilia from *Tetrahymena* by dispersing the cytoplasm with digitonin and purifying the remaining residue by repeated centrifugal fractionation. The drops of purified suspensions were stained and fixed with uranyl acetate, and then critical point dried with carbon dioxide. However, she did not present her illustrations as stereomicrographs, which would have made the interpretation easier. The writer, in early, unpublished experiments, used a similar method for the basal bodies and flagellar apparatus of myxomycete swarm cells. Dissolution with 0.5% digitonin was stopped by adding an equal quantity of dilute osmium tetroxide (0.1 to 0.5% in distilled water) to the partly dissociated cells at the appropriate time as observed by phase microscopy.

Sandborn and Makita (1972) fixed and mechanically dispersed cells on a Smith and Farquhar chopper. Diastase and other unspecified enzymes were used to remove outer membranes. Using critical point drying and stereo observations, these authors showed interconnections of mitochondria with each other and bridges between other organelles. Pihl and Bahr (1970a and b) used novel techniques, and applied quantitative methods to the study of mast cell granules and mitochondria. To avoid lipid loss, cells, either after very brief fixation or with no fixation, were dehydrated with ethylene glycol. For the intermediate fluid, they used ethylene glycol monoethyl ether (Cellosolve) followed by carbon dioxide-critical point drying. The matrix structure visible in critical point dried mitochondria was not apparent in air-dried specimens. Wolfe (1965) examined microtubules isolated from salamander (*Tarista*) erythrocytes and rat sperm tails. After spreading on distilled water, the specimen was picked up on grids and treated as follows: uranyl acetate (2%) 10 min (fixation)/water rinse/ethanol, 50% cold, 70, 95, 100% (3 times)/amyl acetate, 100% (3 times)/CO_2-critical point drying.

Chromosomes and Chromatin. Critical point drying is playing an important role in the developing concepts of chromosome structure. The literature is too vast to be reviewed here; only some recent examples will be mentioned. (See Cohen *et al.*, 1968, for bibliography of earlier literature; Ris and Kubai, 1970, for a comprehensive review; Hayat and Zirkin, 1973, for brief directions for preparation; and Lang, 1972, for simple techniques for spreading chromatin.)

In chromosome studies, the objectives range from attempting to preserve intact metaphase plate to the ultimate unraveling of the DNA fibers of chromosomes. The problems in the preservation of chromosomes individually and in relation to each other are the same as the problems of any organelle isolation—that is, to provide sufficient physical or chemical disruption of the cell to free the organelle but not sufficient to damage the isolated structure. Grinding, chopping, enzymatic, and surfactant treatments are used to achieve this goal. In the examination of chromosome structure or DNA arrangement within the nucleus, the attempt goes further. The structure is partly or totally disrupted so that the nucleic acid fibers become apparent. This is generally accomplished by using surface tension to spread the structures over the surface of a liquid by the Kleinschmidt technique or its variants.

It would seem that the two-dimensional preparation by spreading of DNA strands would not yield any more information by critical point drying than by ordinary air-drying. Nevertheless, shape and relations between fibers are better resolved after critical point drying than after air-drying (Wolfe and Grim, 1967). Comings and Okada (1970) investigated the synaptonemal complex and meiotic chromosome structure of the testes of frogs, mice, quail, and crayfish by procedures which involved mechanical mincing, enzyme treatment, spreading on distilled water, and critical point drying via amyl acetate/carbon dioxide. Their paper has valuable details of the procedure. They later (1971a) extended this work to a study of the kinetochore fine structure in the Indian muntjac, selecting this dwarf deer because it has the smallest number of (and hence rather large) chromosomes of any known vertebrate. Okada and Comings (1970) utilized very small chicken microchromosomes for their attempt to resolve the controversy of how the DNA fibers are arranged in the chromosome. A condensed summary of the fact and opinion has been presented by Comings and Okada (1971b). It is clear that critical point drying alone, while helpful, does not provide the authoritative data for chromosome structure.

In an attempt to study entire metaphase plates by quantitative electron microscopy, Golomb and Bahr (1971a) used a simple procedure of spreading cultured leukocytes on somewhat hypotonic salt solution, picking them up on Formvar-coated grids, and rapidly processing through 30% and higher concentrations of ethanol/amyl acetate/carbon dioxide-critical point drying. After studying 3,900 grids, although many chromosomes were obtained, they found only one entirely visible metaphase plate. Considerably more success was attained by these authors (1971b) in an SEM study of the surface structure of human chromosomes using essentially the same techniques as previously described.

In general, stereoscopy has not been used to take full advantage of critical point drying in chromosome studies (see, however, Ris and Kubai, 1970). By experimenting with various buffer solutions and using salts and urea for selective removal of components, Stubblefield and Wray (1971) prepared numerous excellent stereomicrographs of hamster metaphase chromosomes. They used no spreading techniques, but after treating the specimens as described above, they placed suspensions directly on Formvar-coated grids and processed them by ethanol/amyl acetate/carbon dioxide-critical point drying.

Reproductive Structures. Lung (1968) treated bull and human sperms with thioglycollate to disrupt the tough outer membrane, obtained partial spreading on a Langmuir trough, and processed them by carbon dioxide-critical point drying. Apparently branched and disorderly arrays of fibers were shown in both types of sperm: bull sperm fibers ranging from 14 to 24 nm, and human sperm fibers extending from the same to possibly as small as 7.5 nm. Zirkin, in a later study (1971), found fibers ranging from 16 to 18 nm. The sperms of frog, goldfish, and sea urchin were spread on distilled water followed by uranyl acetate (2%)/

distilled water rinse/ethanol/amyl acetate/carbon dioxide-critical point drying.

Tegner and Epel (1971) studied the fertilization reactions at the surface of sea urchin eggs by intensive sampling during the first three minutes after fertilization. The specimens were processed by 4% glutaraldehyde in seawater diluted to 35% (slightly hypotonic)/ethanol/intermediate fluid not mentioned/Freon (13?)-critical point drying. They showed excellent micrographs, including one at 20,000× of a sperm attached to the egg.

Minute Specimens (Spores, Pollen Grains, Large Cells). The general method of handling these specimens has already been considered under Specimen Containers. Dry spores and pollen grains (pollen grains are microspores, and the distinction is one of convenience) should be wetted so that they can swell to their physiologically active size. They can be fixed either by exposing a suspension in a thin layer to the vapors of osmium tetroxide, as described previously, or by mixing the suspension with an equal quantity of 2% osmium tetroxide in distilled water or dilute (0.05%) buffer. The solutions can hardly be too hypotonic, as the cell wall will constrain excessive swelling; but if hypertonic, there is danger that wrinkles and dents will remain and the spores may not reach fullest size. The critical point dried myxomycete spores (Fig. 2–9) and willow pollen grains (Figs. 2–7 and 2–15) were treated in this manner. They were prepared after fixation by water/ethanol, continuous gradient/Freon TF, continuous gradient/Freon 13-critical point drying.

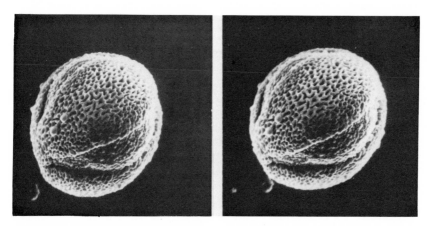

Fig. 2–15. A stereopair of pollen of willow (*Salix scouleriana*). 1,500 X.

Moist or wet specimens can be fixed by immersion in 0.5 to 1% osmium tetroxide for 1/2 to 1 hr. Spirogyra cells thus prepared showed quite acceptable preservation of ultrastructure with 1% osmium tetroxide unbuffered and at room temperature (Cohen and Shaykh, 1973).

Cultures on Agar. Using glutaraldehyde fixation, Bibel and Lawson (1972) found Freon-critical point drying the most reliable method for bacterial colonies. The following procedures have been developed in the author's laboratory. The procedures are short and simple; much of the discussion is concerned with avoiding mistakes.

Cultures on agar can be treated as large specimens with a few additional precautions. Although it is possible to fix an entire agar plate and then select regions, preferably regions should be cut out before fixation. In the latter case, layers (2 to 3 mm) of agar are easier to process than are thick layers. Two % agar properly dehydrated and critical point dried shows no discernible shrinkage. Therefore the flat, sharply defined surfaces of agar blocks are maintained during processing. When possible, select young colonies of bacteria, molds, or other microorganisms; fuzzy, slimy, or overgrown cultures should be avoided.

Aqueous agar gel is likely to exude drops of water when compressed (syneresis), which can detach the colonies. Therefore bending or compression should be avoided. With a sharp razor blade, make vertical cuts completely through the agar. Wet the razor blade and slide it through one of the cuts to lift the desired block. It may be necessary to trim away some of the surrounding agar in order to lift the block without compressing it. Add drops of water as necessary so that capillarity draws them under the block; the sample can then be easily lifted out on a film of water and transferred to a slide or dish without distortion or breakage.

Expose the block to the vapors of osmium tetroxide (1 to 4%) for approximately 10 min or until the surface is distinctly brown. It may then be examined under a dissecting microscope or low power of the compound microscope for particular areas, and these are marked. This is accomplished by trimming the block to make it irregular and asymmetric and making an outline sketch of it. The sketch serves to indicate which is the upper side and to map interesting localities. Mark these on the sketch.

Mold colonies, with aerial hyphal that are difficult to wet, may be osmium vapor fixed and then introduced directly into 30% ethanol, face up. Other types of colonies particularly those which are more easily wetted or which are not anchored to the agar by penetrating hyphae should be handled differently. If they are slipped under the surface of water or fixative, there is a danger that most or all of the surface growth will detach as a film. Therefore lift the block by sliding the wet blade under it, turn it upside down (if not much larger than the area of contact with the blade, the block will adhere by capillarity), and lower it face down and horizontally into a dish of water or buffer. The depth of liquid should be sufficient to allow the block to be turned right side up without breaking the surface. The block will detach and drift gently down. Without letting it scrape the bottom, carefully turn it over. The liquid may be substituted by buffered or unbuffered osmium tetroxide, glutaraldehyde (1%), or formaldehyde, and, after 0.5 to 1 hr, rinsed and dehydrated. After vapor fixation, additional fixatives harden the colonies and thereby facilitate sectioning for TEM studies.

These initial aqueous changes are easily made either by pipetting off the old liquid and replacing it or by lifting the block horizontally so that a layer of water remains on the surface. The folded stainless steel screen holders (Fig. 2–4b) make excellent carriers for agar blocks, and serve to hold them under the dense intermediate fluid (Freon TF), in which they would otherwise float. If desired, agar can be dehydrated by rather large steps, as agar and most of the microorganisms on its surface are not very sensitive to concentration changes. The following schedule is suggested. The specimens are exposed to osmium tetroxide (4%) vapors for 1 hr, rinsed, dehydrated with 30, 50, 70, 90, 95, and 100% (twice) ethanol, substituted with 10, 25, 50, 75, and 100% (twice) Freon TF, and then Freon 13-critical point dried. The length of time in each substituent fluid depends upon the thickness of the block; for a 2 mm thick block with all surfaces exposed for exchange, 10 min is sufficient. It is essential that all traces of water or alcohol be removed before critical point drying, and pressure should be relieved gradually. If the block is not completely dehydrated and alcohol-free, it will collapse and distort later. If pressure release is too fast, bubbles form and distort the agar. If these faults are avoided, the block retains its shape and dimensions. The difficulties reported in the literature with critical point drying of agar are probably caused by failure in these regards.

The dry agar blocks are extremely delicate and light, as 2% agar has only one-fiftieth of its original weight after drying. They should either be lifted on a blade or, if held with forceps, be grasped at a part which will not be used for examination. Blocks should be affixed to stubs with tacky but not wet adhesive, as absorption of liquid is ruinous. Double-coated Scotch tape or a very thick conductive cement troweled into a smooth layer makes a satisfactory adhesive. The myxamebae, spores, and adventitious bacteria in Figs. 2–16 and 2–17 were prepared by preliminary vapor fixation of agar cultures followed by inverted rinse in

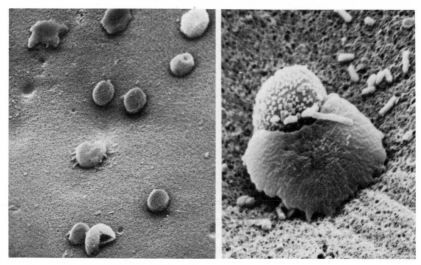

Fig. 2–16   **Fig. 2–17**

Figs. 2–16 and 2–17. Spores and myxamebae of *Didymium iridis* (Myxomycete) with adventitious bacteria. Fig. 2–16 shows a group of spores (rounded bodies) and myxamebae (irregular bodies) on agar. Note empty germinated spore case at bottom. Also, note a depression of agar near the center; this is a frequent occurrence if the beam scans a small area on agar. 850 X. Fig. 2–17 shows a myxameba partly surrounding a spore. Note rod-shaped bacteria on the spore, ameba, and agar. Depression of agar around spore and ameba is produced by the scanning beam. 3,000 X.

water, postfixation with 1% osmium tetroxide in distilled water, and continuous gradient fluid substitution.

Higher Plants. Cohen and Garner (1971) have used osmium tetroxide vapor fixation routinely for the fixation of higher plants (roots, root hairs, small stems, leaves, and flowers). Use fresh, healthy specimens. If there is a choice, select small leaves, flowers, or even entire small plants. Before fixation, specimens may have to be cleaned by gentle rinsing, brushing, or puffs of air. Restore normal turgor, if desired, by keeping in a moist chamber (e.g., a covered dish lined with wet paper towel or filter paper). Small structures (e.g., flowers and leaves having less than 1 cm diameter) should remain attached to stems to provide convenient handles and to preserve natural relationships; large ones should be cut into portions which will loosely fit the smallest chamber used in processing.

In any convenient shallow container with a lid (e.g., petri dish), place the plant specimens to be fixed, a small open dish containing 2 to 4% osmium tetroxide, and replace the lid. The specimens may be left until they have attained maximum blackness (usually 3 to 4 hr) or overnight. Although such prolonged fixation seems useful for penetration into leaves and small stems, some flower petals wilt, and root hairs may collapse after several hours. Therefore a trial processing is desirable.

After vapor fixation, some further trimming can be carried out rapidly with fine scissors and the specimens placed in baskets and dehydrated by closely spaced steps (or continuous gradient). The dehydration is carried out with ethanol (20, 30, 40, 50, 60, 70, 80, 90, 95, and 100% [twice]), substituted by Freon TF (10, 20, 30, 45, 60, 75, 85, 95, and 100% [twice]), and Freon 13-critical point dried. Most plant structures exchange fluids rapidly after any external waxy layers have been penetrated, and dehydration may be relatively rapid for small specimens (approximately 5 min at each step).

The use of containers for small plant parts is essential, for these organs become extremely brittle on dehydration. Very delicate structures, such as root hairs, cannot withstand repeated removal and immersion in solutions. After a brief vapor fixation, the roots should be put into water containing a few drops of detergent to insure wetting and kept completely under fluid until the process is finished. The container illustrated in Fig. 2–5c is designed for this purpose.

Animal Tissue Culture. Tissue cultures are usually a thin layer of cells attached to a substrate, and need no preliminary special handling except trimming, and they can be passed through the fluid substitution series rapidly. However, these specimens are more subject to osmotic stresses than are massive tissues, because the cells are almost as exposed to the action of fixative as are single cells. Overman (1963) grew chicken fibroblast cells directly on coated grids, fixed the cultures with veronal-buffered osmium tetroxide, and processed through ethanol/amyl acetate/carbon dioxide-critical point drying. The specimens were lightly shadowed for study with the TEM.

Turnbill and Philpott (1970) were probably the first to publish application of Freon-critical point drying to animal tissues. They reported briefly on the preparation of fibroblasts, entire mouse heart, and starfish tube feet; the fixative was not specified. Sträuli and Haemmerli (1971) used Parducz's fixative for tissue cultures to demonstrate surface specializations of cells in tissue culture. These authors specifically advised against the use of aldehydes for fixation. Price (1972) studied the membranous "ruffles" of cultured monkey kidney cells and their relation to the cell. He published excellent micrographs of the surface examined with the SEM and micrographs of sections examined with the TEM. Based on these micrographs, he constructed wax models of the cell structure. This procedure was applied to cultures on coverslips, and consisted of fixation with glutaraldehyde (3%) followed by osmium tetroxide (1%), rinse with a buffer, dehydration in ethanol (20, 40, 70, and 100%), substitution with amyl acetate, and carbon dioxide-critical point drying.

Boyde and Vesely (1972) examined tissue cultures expressly for a comparison of fixation and drying procedures. They concluded that glutaraldehyde fixation was, in general, superior to osmium tetroxide fixation for the SEM examination, that there was no difference between carbon dioxide- and Freon-critical point drying, and that N_2O-critical point drying was unsatisfactory. This writer suggests the following procedure for the preservation of fine surface features.

The specimens are fixed for 1 hr with 1% osmium tetroxide buffered with cacodylate (pH 7.0) and isotonic or slightly hypotonic with the culture medium. They are then passed through closely graded dehydration, intermediate fluid exchange, and critical point dried. For combined SEM and TEM studies, which require the preservation of internal structure, the procedure presented by Price (1972) is recommended.

Perecko et al. (1973), in a study of the effect of oncogenic viruses on cells, grew cultures of murine tissues on small (6 mm diameter) coverslips, which were then processed in toto through 3% glutaraldehyde/osmium tetroxide/ethanol/amyl acetate/carbon dioxide-critical point drying. They used carriers similar to those depicted in Fig. 2–5a and b to process six coverslips simultaneously.

Animal Organs and Tissues. The mutual support of cells and the penetration gradient forced by the thickness of most soft animal tissues allow considerable leeway for acceptable results. Boyde (1972) discussed these factors, and suggested glutaraldehyde fixation followed (after thorough washing) by osmium tetroxide postfixation. After a thorough investigation of fixatives, methods of fixation, fluid processing, and final drying, Nowell et al. (1972) concluded that artifact-free surfaces showing fine detail were observed only in specimens fixed by aldehyde perfusion, critical point dried with carbon dioxide, and metal-coated. The fixative used was Karnovsky's mixture of glutaraldehyde and formaldehyde; fixation by bronchiolar perfusion was necessary to preserve the cilia in normal relationship.

For investigating the lactating mammary gland, Nemanic and Pitelka (1971) used 2% unbuffered glutaraldehyde for 24 hr, after which the tissues were sufficiently firm to be sliced easily. They were then treated for 24 hr each with 16% glycerol and 20% ethanol to remove the milk fat globules prior to further processing. The dehydration is carried out with ethanol (50, 75, 95, and 100% [twice]) for 15 min at each step, sub-stituted with Freon TF (50 and 100% [twice]) for 15 min at each step, and Freon 13-critical point dried. Their micrographs showed well-preserved alveoli and microvilli.

In order to study the relationship of intestinal microorganisms to the luminal cells of the intestine, Erlandsen et al. (1972) avoided washing out the intestine; instead, the tissue was immediately fixed and carbon dioxide-critical point dried. *Giardia, Hexamita,* and bacteria were found on the villi and in the crypts, and *Trichomonas* was found on the mucosa of the large intestine. Details of the preparation were not given.

Massive organs (liver, kidney, and testis) are perfused with aldehydes when practicable, then cut to expose inner structure. Fixed soft organs are far easier to dissect cleanly than fresh ones. The intestine and blood vessels in Figs. 2–18 and 2–19 were prepared by first flushing out the intestine with saline, then injecting 1% osmium tetroxide in 0.1 M phos-phate buffer (pH 7.4) into the lumen. The intestine quickly hardened in distended condition. It was removed, placed in additional fixative for one hour, rinsed, and further trimmed and processed. The dehydration

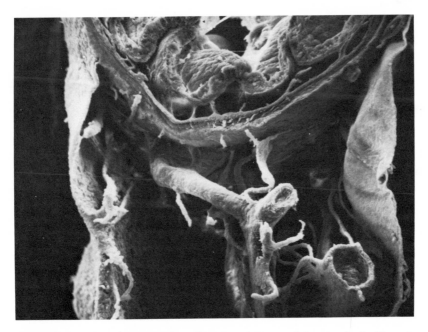

Fig. 2–18. Mesenteries, blood vessels, and mucosa of frog small intestine fixed with osmium tetroxide and critical point dried with Freon. 30 X.

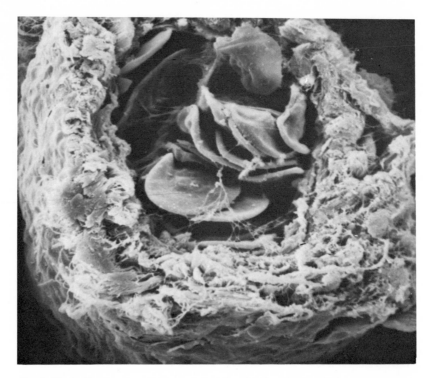

Fig. 2–19. Small artery in mesentery of frog intestine (see Fig 2–18) showing erythrocytes. 2,000 X.

and substitution were carried out with ethanol (continuous gradient) and Freon TF (continuous gradient), and Freon 13-critical point dried.

Osmium tetroxide is preferable to reduce charging artifacts. But glutaraldehyde is a cheaper fixative, not as irritating as osmium tetroxide, maintains approximate natural colors to aid in orientation and dissection, and is easier to use on larger specimens. In the writer's laboratory, osmium tetroxide is the preferred fixative for small specimens, for those in which good fixation of the exposed surface only is important, and for specimens in which the natural color pattern is not necessary for orientation. Glutaraldehyde is preferred for large specimens, for perfusion (carried out wherever possible), and where the specimen will later undergo detailed dissection.

Osmolarity, pH, temperature, and duration of fixation become critically important when the objective is to preserve minute details of the cell interior or surface and to avoid shrinkage of cells and tissues away from each other.

Inner Structures of Tissues and Cells. The study of cell surfaces and their spatial relationships has been enhanced immeasurably with the aid of SEM. For the first time, the biologist can see lung alveoli, the organ of Corti, synaptic junctions, mesophyll, and palisade layers of leaves as they actually are, not as reconstructed in drawings or models.

For animal tissues, clean, sharp cuts made through the tissue before or after fixation is often all that is needed to expose layers of tissues and cavities. Plant tissues are generally brittle after critical point drying, and excellent preparations can be made simply by snapping off small organs. Figs. 2–20 and 2–21 of oregano leaf were prepared in this manner.

Specimens can also be frozen in liquid N_2 before or after critical point drying, which temporarily increases their brittleness and makes them more easily fracturable. Nemanic (1972) has described some ingenious methods for this purpose. For very small or microscopic specimens, other methods must be used. Tanaka *et al.* (1972, 1973) used a resin cracking method, which is presented in detail in this volume. Lim (1971) dehydrated inner ear, liver, kidney, and "house dust mites" with 70% ethanol, then dropped minute pieces of these specimens in drops of alcohol on aluminum foil cooled to liquid N_2 temperature. The drops froze and spontaneously cracked through the specimen. Although subsequent treatment was for sectioning and freeze-dehydration, the method is adaptable for critical point drying.

For the purpose of protecting delicate structures from collapse before vacuum dehydration, Otaka and Honjo (1972) cooled the specimens in 100% ethanol to liquid nitrogen temperatures. As it is cooled, ethanol becomes increasingly viscous and finally hardens to a brittle glass and therefore does no damage by the formation of crystals which pierce the cells. This writer adapted this technique for the examination of filamentous algal cells (*Spirogyra*) by freezing bundles of filaments and breaking them with a razor blade or with a sharp chisel tapped with a hammer. It is essential that all surfaces and instruments which come in contact with the frozen specimens be at liquid nitrogen temperature; otherwise, instead of a fractured block, one has a syrupy puddle. After fracturing, specimens were placed in 100% ethanol for further processing.

Cryofractioning after critical point drying was used by Sybers and Ashraf (1973) in a study of cardiac muscle. Although presumably water-free, the muscle cooled to liquid nitrogen temperature became brittle and fractured easily.

Combined modes of observation on one specimen can frequently provide considerably more information than would be expected by the mere summation of data obtained from each. Thus Wetzel *et al.* (1973)

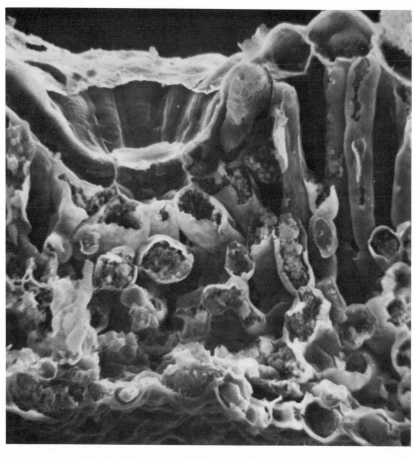

made photomicrographs of the leukocytes in moist Giemsa-stained blood smears, processed these smears through critical point drying, and made SEM micrographs of the same cells, thereby correlating the external morphology of leukocytes with their well-known appearance under the light microscope.

In a combination of optical microscopy, TEM, and SEM, Pachter *et al.* (1973) made 20 μm thick serial sections of lung and muscle and, after preliminary survey of these sections, remounted and resectioned them at 1 μm for optical micrography and SEM studies and prepared "ultrathin" sections for the TEM. The epoxy embedding resin was removed from the sections by prolonged (1 to 2 weeks) etching with a solution of iodine in acetone.

Wickham and Worthen (1973) obtained very clear TEM micrographs of retina and other ocular tissues from sections of tissues embedded in epoxy resin after critical point drying and after SEM examination. Wickham discusses the procedure in detail in volume 2 of this series. In an alternative procedure, Erlandsen *et al.* (1973) first sectioned conventionally fixed, dehydrated, and embedded tissues for TEM study, then removed the epoxy resin from the block with sodium ethylate and examined the cut surface in the SEM.

Although the examination with the SEM has made the interrelationships of cells directly visible in a dramatic and informative manner, it has not been as informative in regards to the internal cell structure. More information can be obtained from TEM examination of sections or freeze-etch replicas. The unsatisfactory appearance of fractured cells is probably due to the numerous partly transparent and confusingly overlapping shreds of cytoplasm and to charging phenomena. The extremely irregular surfaces of fractured cells make poor conducting pathways. New and improved techniques, such as these described, are mastering these obstacles.

STORAGE, MOUNTING, COATING, AND EXAMINATION OF SPECIMENS AFTER CRITICAL POINT DRYING

In a critical point dried specimen, all of the original liquid and soluble solid components (e.g., salts and fats) are lost, and the space they oc-

Figs. 2–20 (top) and 2–21 (bottom). Leaf of oregano (*Origanum vulgare*) fractured after critical point drying. Fig. 2–20 (top) shows a glandular pit from which the drop of secretion has been lost. The upper palisade cells and the lower interlaced spongy mesophyl cells are prominent. 700 X. Fig. 2–21, another portion of oregano leaf, showing upper epidermal surface and a pit gland with a drop of secretion. 450 X.

cupied is replaced by air. Therefore some critical point dried specimens are exceedingly light, soft, or brittle. When possible, pick up specimens by scooping them up on spatulas or knife blades; otherwise, handle them carefully with forceps, holding the specimens preferably by parts which will not be studied. Animal tissues and agar substrates of cultures crush easily, while plant tissues such as fine stems, leaves, and flowers are often extremely brittle.

Unless they are to be mounted immediately, specimens should be placed in a desiccator as a protection against moisture and dust. Excessive desiccation or vacuum is not necessary. Silica gel containing a color-changing moisture indicator is the recommended desiccant. Its condition can be easily observed, and it is inexpensive and can be regenerated in an oven at 60 to 80°. Small screw-capped jars may be used as desiccators by placing a little silica gel in the bottom of the jars and covering the desiccant with filter paper. Do not crowd specimens, as extricating one from a mass of tangled or packed objects is almost certain to cause damage. Very minute specimens (pollen and spores) may be left on their original filters until needed. Small plastic petri dishes are satisfactory containers for such specimens.

Specimen Mounting

The conventional mountant for specimens is a pressure-sensitive adhesive tape coated on both sides ("double-stick Scotch tape"). The major advantage of this mountant for most specimens is convenience; otherwise it is usually inferior to some others (Leffingwell and Hodgkin, 1971; Johari and DeNee, 1972). Leffingwell and Hodgkin (1971), in an exhaustive study of various mountants, concluded that a polyvinyl chloride (PVC) solution (International Coatings Co., 5V-229 Cl; see Appendix for sources) was the best from the standpoint of damage by the scanning beam, sinking of small specimens into the adhesive, and contamination. This writer has also observed less charging with PVC than with the tape. Polyvinyl chloride has an additional advantage of providing dark, even background in the SEM.

Double-faced adhesive tape is still useful for attaching large, flat specimens (e.g., circles of membrane filter bearing spores or other small particles), difficult to attach otherwise, and under those circumstances where adhesive is not exposed to the beam. The tape should not extend beyond the border of the stub, and a dab of conductive cement should be applied to electrically connect the specimen to the metal of the stub. Horridge and Tamm (1969) coated stubs with a dilute chloroform solution of the adhesive from cellophane tape and sprinkled their minute specimens on the dried adhesive.

The following directions are for polyvinyl chloride (PVC) cement: Dilute approximately 5 ml of the stock supply with an approximately equal quantity of methyl ethyl ketone. Stirring, ultrasonic treatment, or cautious warming to approximately 50° will speed the mixing of the gelatinous stock with the solvent. The final consistency of PVC should be semifluid (similar to thick, raw egg white). Spread the PVC over the surface with a glass rod or wooden applicator in a smooth puddle and allow it to dry to tackiness. Although a thin solution of PVC is much easier to spread, it does not seem to dry to the desirable adhesiveness.

Spread particles (e.g., pollen grain and fungus spores) lightly over the surface by gently tapping the filter paper above the stub or by touching the cotton tip of an applicator to the particle mass and gently tapping the applicator stick with the finger while it is held close above the stub. Sometimes the coated face of the stub can be touched to the particle mass. It is better to have a light layer of particles so that they be separated from one another. Set the particles into the cement by very gently directing a puff of air from an ear syringe against the stub, or use a cautious puff of gas from a duster can. The breath, because of its humidity, should not be used.

The PVC cement can be used to hold minute insects or other small specimens. Specimens which are too small to be picked up with forceps or are too large to be dusted on the surface present a difficult problem. They can be picked up, depending upon the size and shape, on the tip of a fine needle barely touched with PVC which is allowed to dry, on a small watercolor brush which has been cut down to one or two bristles, or on a toothpick dry or tacky with PVC. They should never come in contact with any free liquid.

Larger specimens with an irregular surface (e.g., flowers, leaves, and rings of intestine) are best fastened down with a conducting cement. These cements vary in suitability, but two very suitable kinds are available (see Appendix). One of these is very fluid, and the other is a thick paste. The fluid one, after some drying on the stub, is suitable for specimens with an extended base or irregular contact surface along which the fluid can creep and establish satisfactory contact. The paste is better for small macroscopic specimens (1 mm or larger) of very irregular shape. Sometimes it is advantageous to cover a dab of thick silver paste on the stub with a drop of the thin conducting cement to get intermediate consistency.

Push the uninteresting end of the specimen into the cement, orient, and hold for a moment until the paste sets. Insects oriented for desired view may be gently pressed on the stub. If a dab of thick paste on a toothpick is touched to a stub and the toothpick pulled away, a soft cone is left. Single small specimens may be very precisely mounted on these cones by first (under a dissecting microscope) picking up the specimen with a smaller cone on the paste-covered tip of a toothpick.

The liquid should not be used with porous material (e.g., dried agar and membrane filter) in which the solvent can be drawn up by capillarity. Agar collapses immediately, and the exposed face of the membrane becomes covered with a film of the cement. For such specimens, use the paste after most of the solvent has evaporated, PVC, or, if suitable, the double-faced adhesive tape.

Very Minute Specimens. Exceedingly small specimens of 50μm or smaller in size (e.g., pollen grains, cell organelles, and bacteria) which have been carried through the fluid replacement series on a membrane filter, agar surface, grid, or other surface, and which are not excessively crowded, need not be transferred to another mount or adhesive. The original substrate may be fastened to a stub in preparation for coating. Johari and DeNee (1972) describe a variety of ingenious mounting procedures for small particles.

Specimen Coating and Examination

Since critical point dried specimens are particularly subject to charging, satisfactory coating is of primary importance. A double coating, first of carbon followed by a metal, is preferred over the single metallic coating. Carbon, after evaporation, tends to deposit in shadowed recesses not necessarily exposed to the source. Therefore a carbon coat will insure electrical conductivity across those areas not covered by the succeeding metal. For the metal layer we routinely use gold; however, Dr. Norman Hodgkin (unpublished communication) recommends an alloy of 60% gold and 40% palladium. The Au-Pd mixture provides a continuous layer with a much thinner deposit than does pure gold, and is less likely to crack under the beam or on storage.

No quantitative criteria for coating thickness will be given here, as the type, size, and shape of the specimen and conditions of examination (accelerating voltage and magnification) are variable conditions for a problem whose best solution is learned by experience. The following suggestions are intended as guidelines:

(1) Surfaces to be scanned should be as close to the stub as possible in order to decrease the length of the conductive pathway. Blocks of

tissue should be thin slabs attached by a broad face. Tubular organs set on end should be more like rings than chimneys.

(2) Small specimens, which are dusted or sprinkled on the surface, should not be allowed to crowd; the particles should not shade each other from exposure to the evaporated metal.

(3) Carbon should be deposited as slowly as possible without excessive heating. The latter is a real danger, for the carbon rods are white hot and radiate sufficient heat to damage specimens on prolonged exposure. A vacuum of 10^{-4} to 5×10^{-6} torr is considered desirable, although this point has not been extensively investigated for SEM specimens (Echlin and Hyde, 1972). A lower (poorer) vacuum yields a more diffuse carbon coat which can layer undercut surfaces, whereas a higher and "cleaner" vacuum (i.e., without oil vapors) insures better surface contact. For metal deposition, the higher the vacuum the better the results are.

(4) Gold or other metal should be deposited slowly on the revolving specimens to insure thorough and even coating. The relatively low temperature and small area of the evaporation source do not endanger the specimens as much as does exposure to carbon evaporation. On completion, glossy smooth surfaces (i.e., the metal of the stub and exposed PVC cement) should be golden color; matte (membrane filter, agar, and some specimens) and rough surfaces may be gray to green. If the coating is seen to be insufficient, the specimen may be recoated. Large specimens (organs and tissues) should be coated quite heavily; the amount of coating is dependent upon the degree of details necessary to be examined. There is always a possibility of obscuring very fine surface details as a result of excessive coating or, on the other hand, of having charging artifacts because of inadequate coating. Minute specimens (bacteria and cell organelles), which are close to the stub and have a short conduction path, should be coated lightly to preserve the details.

(5) Examine the specimens as soon after coating as possible, and keep them desiccated and shielded from dust or vapors. If the examination is not completed within a week, the coatings may crack, especially if they are exposed to varying conditions of temperature, humidity, and alternate vacuum and pressure. Lewis and Nemanic (1973) recommend immediate coating after critical point drying to obtain good adhesion and electrical conductance. While we concur if specimens are going to be examined immediately, our experience indicates that it is better otherwise to delay coating until just before examination. If charging is intractable, recoat the specimens.

(6) Use low beam voltage (10 kV instead of 20 kV) to minimize

charging and damage when possible. Avoid unnecessary exposure to the beam; focus, take the picture rapidly, and then turn off the beam when the specimen is not being examined.

(7) In those cases in which charging persists, especially those involving spores or other small particles, change to another field, or admit air to the chamber and pump down again.

(8) It is also desirable to use backscattered instead of secondary electrons or to expose the specimen through a small hole in metal foil cover (Pfefferkorn *et al.*, 1972). Devices for reducing charging artifacts by altered chemical and electrical environments of the specimen have been described (Garner *et al.*, 1973), but these have not been subjected to wide use and evaluation.

The normal three-dimensional structures may be exhibited in very informative relief by shadowing (TEM) and stereomicrography (TEM and SEM), as shown in Figs. 2–14 and 2–15 (See Nemanic in this volume for details.)

The writer is indebted to Ms. Joan Sims, Mr. Greg Carras, and Mr. R. G. E. Steever for technical assistance, and to Mrs. Candess Eliason for typing the manuscript and preparing the list of references. Mr. Dan Marlow and Mr. Gerald Garner read the manuscript critically and made valuable suggestions. The work in the Electron Microscope Center was partly supported by a grant from the Research Corporation and by equipment grants from the Biomedical Research Funds of the state of Washington and the National Science Foundation.

APPENDIX

Materials

The Ted Pella Company is the source for the liquid and paste silver conducting paints, and polyvinyl chloride cement (trade name, Mikrostik). They are the American agents for the Polaron critical point drying devices. Ted Pella Company, P. O. Box 510, Tustin, California 92680.

The Du Pont Company is the source for the various fluorocarbons (Du Pont trade name, Freons) and for general information. Freon 13 is often obtainable from refrigeration supplies. Oil-free Freon should be specified. E. I. du Pont de Nemours & Company, Inc., Freon Products Division, 701 Welch Road, Palo Alto, California 94304.

The Bomar Company offers various containers for grids, small specimens, filters and cover slips to fit into critical point drying devices. They also manufacture several critical point drying instruments. The Bomar Company, P. O. Box 225, Tacoma, Washington 98401.

Critical Point Drying Apparatus and Devices

Bomar Company. SPC 900, 900 E/X, 50, 50 E/X CPD instruments. Address given above.

Denton Vacuum, Inc. DCP-1 Critical Point Dryer. Denton Vacuum, Inc., Cherry Hill Industrial Center, Cherry Hill, N.J. 08003.

Micrographics. NCT line of critical point drying equipment. Micrographics, 3855 Birch Street, Newport Beach, California 92660.

Parr Instrument Company. Parr Critical Point Drying Apparatus. Parr Instrument Company, 211 Fifty-Third Street, Moline, Illinois 61265.

Polaron Equipment Limited. Polaron Critical Point Apparatus. Polaron Equipment Limited, 60/62 Greenhill Crescent, Holywell Industrial Estate, Watford, Hertfordshire, England. (Ted Pella Company, American agents.)

Ivan Sorvall Inc. Sorvall Critical Point Drying System. Ivan Sorvall Inc., 127 North Bayshore Boulevard, San Mateo, California 94401.

REFERENCES

Anderson, T. F. (1951). Techniques for the preservation of three-dimensional structure in preparing specimens for the electron microscope. *Tran. N.Y. Acad. Sci.* **13**, 130.

Anderson, T. F. (1956). Electron microscopy of microorganisms. In: *Physical Techniques in Biological Research* (Oster, G., and Pollister, A. W., eds.), Vol. III, pp. 177–240. Academic Press, New York.

Anderson, T. F. (1966). Electron microscopy of microorganisms. In: *Physical Techniques in Biological Research* (Pollister, A. W., ed.), 2d ed., Vol. IIIA, pp. 319–87. Academic Press, New York.

Argetsinger, J. (1964). The isolation of ciliary basal bodies (kinetosomes) from *Tetrahymena pyriformis. J. Cell Biol.* **24**, 154.

Arnold, J. D., Berger, A. E., and Allison, O. L. (1971). Some problems of fixation of selected biological samples for S.E.M. examination. In: *Scanning Electron Microscopy/1971* (Johari, O., ed.), pp. 249–56. IITRI, Chicago.

Bessis, M., and Weed, R. I. (1972). Preparation of red blood cells (RBC) for SEM: A survey of various artifacts. In: *Scanning Electron Microscopy/1972* (Johari, O., ed.), pp. 289–96. IITRI, Chicago.

Bibel, D. J., and Lawson, J. W. (1972). Scanning electron microscopy of L-phase streptococci. I. Development of techniques. *J. Microscopy* **95**, 453.

Blinder, S. M. (1969). *Advanced Physical Chemistry.* Macmillan, London.

Boyde, A. (1972). Biological specimen preparation for the scanning electron microscope: An overview. In: *Scanning Electron Microscopy/1972* (Johari, O., ed.), pp. 257–64. IITRI, Chicago.

Boyde, A., and Vesely, P. (1972). Comparison of fixation and drying procedures for preparation of some cultured cell lines for examination in the SEM. In: *Scanning Electron Microscopy/1972* (Johari, O., ed.), pp. 265–72. IITRI, Chicago.

Boyde, A., and Wood, C. (1969). Preparation of animal tissues for surface-scanning electron microscopy. *J. Microscopy* **90**, 221.

Bradley, D. E. (1965). Replica and shadowing techniques. In: *Techniques for Electron Microscopy* (Kay, D., ed.), 2d ed., pp. 96–152. Blackwell Sci. Publications, Oxford.

Bystricky, V., Fromme, H. G., Pfautsch, M., and Pfefferkorn, G. (1972). Scanning and transmission electron microscopy of an unusual soil bacterium: Application of the critical point drying method. *Micron* **3**, 474.

Clayton, J. W., Jr. (1967). Fluorocarbon toxicity and biological action. In: *Fluorine Chemistry Reviews*, Vol. 1 (Tarrant, P., ed.), pp. 201–25, 251–52. Marcel Dekker, Inc., New York.

Cohen, A. L., and Garner, G. E. (1971). Delicate botanical specimens preserved for scanning electron microscopy by critical point drying. *Proc. 29th Ann. Meet. Electron Micros. Soc. Amer.*, pp. 450–51. Claitor's Publishing Div., Baton Rouge, La.

Cohen, A. L., Marlow, D. P., and Garner, G. E. (1968). A rapid critical point method using fluorocarbons ("Freons") as intermediate and transitional fluids. *J. Microscopie* **7**, 331.

Cohen, A. L., and Shaykh, M. (1973). Fixation and dehydration for surface structure preservation in critical point drying of plant material. *Proc. 6th Ann. SEM Symp.*, IITRI, Chicago, pp. 371–78.

Comings, D., and Okada, T. (1970). Whole mount electron microscopy of meiotic chromosomes and the synaptonemal complex. *Chromosoma* **30**, 269.

Comings, D., and Okada, T. (1971a). Fine structure of kinetochore in Indian muntjac. *Exp. Cell. Res.* **67**, 97.

Comings, D., and Okada, T. (1971b). Chromosome Structure. *Proc. 29th Ann. EMSA Mtg.*, pp. 520–21. Claitor's Publishing Div., Baton Rouge, La.

Daniel, R. S. (1973). Scanning electron microscopy studies of intracellular symbiosis of bacteroids and blattids. In: *Scanning Electron Microscopy/1973* (Johari, O., ed.), pp. 489–96. IITRI, Chicago.

DeBault, L. E. (1973). A critical point drying technique for SEM of tissue culture cells grown on plastic substratum. In: *Scanning Electron Microscopy/ 1973* (Johari, O., ed.), pp. 317–24. IITRI, Chicago.

Deichmann, W. B., and Gerarde, H. W. (1969). *Toxicology of Drugs and Chemicals*. Academic Press, New York.

Deter, R. L. (1973). Electron microscopic evaluation of subcellular fractions obtained by ultracentrifugation. In: *Principles and Techniques of Electron Microscopy: Biological Applications*, Vol. 3 (Hayat, M. A., ed.). Van Nostrand Reinhold Company, New York and London.

Dubochet, J., and Kellenberger, E. (1972). Selective adsorption of particles to the supporting film and its consequences on particle counts in electron microscopy. *Microscopica Acta* **72**, 119.

Du Pont (1971). *Freon Product Information*. Bulletin B-2. E. I. Du Pont de Nemours Company.

Echlin, P., and Hyde, P. J. W. (1972). The rationale and mode of application of thin films to non-conducting materials. *Proc. 5th Ann. SEM Symp.*, IITRI, Chicago, pp. 137–46.

Echlin, P., Paden, R., Dronzek, B., and Wayte, R. (1970). Scanning electron microscopy of labile biological material maintained under controlled conditions. *Proc. 3rd Ann. SEM Symp.*, IITRI, Chicago, pp. 49–56.

Enlander, D., Everhart, T. E., Scott, T., Hoo, R., and Drew, L. (1973). The cytopathic effect of herpes simplex virus in cell cultures. In: *Scanning Electron Microscopy/1973* (Johari, O., ed.), pp. 505–12. IITRI, Chicago.

Erlandsen, S. L., Thomas, A., and Wendelschafer, G. (1973). A simple technique for correlating SEM with TEM on biological tissue originally embedded in epoxy resin for TEM. In: *Scanning Electron Microscopy/1973* (Johari, O., ed.), pp. 349–56. IITRI, Chicago.

Erlandsen, S. L., Wendelschafer, G., and Rolston, J. L. (1972). Visualization of intestinal microorganisms by scanning electron microscopy. *J. Cell Biol.* **55,** 71a.

Falk, R. H., Gifford, E. M., Jr., and Cutter, E. G. (1971). The effect of various fixation schedules on the scanning electron microscopic image of *Tropaeolum majus. Amer. J. Bot.* **58,** 676.

Fassett, D. W. (1963). Esters. I. General considerations. In: *Industrial Hygiene and Toxicology,* Vol. II (Fassett, D. W., and Irish, D. D., eds.), 2d rev. ed., pp. 1847–1934. Wiley Interscience Publishers, New York.

Fromme, H. G., Pfautsch, M., Pfefferkorn, G., and Bystricky, V. (1972). Die "Kritische Punkt"—Trocknung als Präparationsmethode für die Raster-Elektronenmikroskopie. *Microscopica Acta* **73,** 29.

Fullam, E. F. (1972). A closed wet cell for the electron microscope. *Rev. Sci. Instr.* **43,** 245.

Garner, G. E., and Bryant, V. M. (1973). Preparation of modern palynomorphs for scanning electron microscopy by the critical point drying method. *Geosciences and Man.* **7,** 83.

Garner, G. E., Cohen, A. L., and Steever, R. G. E., Jr. (1973). An integrated device for controlling charging artifacts in the SEM. *Proc. 6th Ann. SEM Symp.*, IITRI, Chicago, pp. 189–96.

Gleason, M. N., Gosselin, R. E., Hodge, H. C., and Smith, R. P. (1969). *Clinical Toxicology of Commercial Products. Acute Poisoning.* 3d ed., p. 72. Williams & Wilkins Co., Baltimore.

Golomb, H., and Bahr, G. F. (1971a). Analysis of an isolated metaphase plate by quantitative electron microscopy. *Exp. Cell Res.* **68,** 65.

Golomb, H., and Bahr, G. F. (1971b). Scanning electron microscope observations of surface structure of isolated human chromosomes. *Science* **171,** 1024.

Handbook of Physics and Chemistry (Weast, R. C., ed.), Chem. Rubber Company, Cleveland, Ohio, 50th ed.

Hayat, M. A. (1970). *Principles and Techniques of Electron Microscopy: Biological Applications,* Vol. 1. Van Nostrand Reinhold Company, New York and London.

Hayat, M. A., and Giaquinta, R. (1970). Vapor fixation prior to fixation by immersion for electron microscopy. In: *Microscopie Electronique*, Vol. I, pp. 391–92. 7th Intern. Cong. EM, Grenoble, 1970.

Hayat, M. A., and Zirkin, B. (1973). Critical point drying. In: *Principles and Techniques of Electron Microscopy: Biological Applications*, Vol. 3 (Hayat, M. A., ed.). Van Nostrand Reinhold Company, New York and London.

Heslop-Harrison, Y. (1970). Scanning electron microscopy of fresh leaves of Pinguicula. *Science* **167**, 172.

Horridge, G., and Tamm, S. (1969). Critical point drying for scanning electron microscopic study of ciliary motion. *Science* **163**, 817.

Idle, D. B. (1971). Preparation of plant material for scanning electron microscopy. *J. Microscopy* **93**, 77.

Jard, S., Bourguet, J., Carasso, N., and Favard, P. (1966). Action de divers fixateurs sur la perméabilité et l'ultrastructure de la vessie de Grenouille. *J. Microscopie* **5**, 31.

Johari, O., and DeNee, P. (1972). Handling, mounting, and examination of particles for scanning electron microscopy. *Proc. 5th Ann. SEM Symp.*, IITRI, Chicago, pp. 249–56.

Kanazawa, K., Hamano, M., and Akahori, H. (1972). Application of scanning electron microscope to gastrointestinal research. *Jap. J. Clin. Electron Micros.* **5**, 307.

Kimball, G. (1951). The liquid state. In: *Treatise on Physical Chemistry* (Taylor, H. and Glasstone, S., eds.), Vol. II, pp. 353–419. Van Nostrand Company, New York.

Koller, T., and Bernhard, W. (1964). Séchage de tissus au protoxyde d'azote (N_2O) et coupe ultrafine sans matière d'inclusion. *J. Microscopie* **3**, 589.

Kuwabara, T. (1969). The scanning electron microscopic study of the cell surface. *Proc. 27th Ann. EMSA Mtg., St. Paul, Minn.*, pp. 36–37. Claitor's Publishing Div., Baton Rouge, La.

Kuwabara, T. (1970). Surface structure of the eye tissue. *Proc. 3rd Ann. SEM Symp.*, IITRI, Chicago, pp. 185–92.

Landboe-Christiansen, E., and Parapat, S. (1972). The gastrointestinal mucosa in man, its surface architecture: Some observations by the scanning electron microscope. *JEOL News* **9e**, 12.

Lane, W. C. (1970). The environmental control stage. *Proc. 3rd Ann. SEM Symp.*, IITRI, Chicago, pp. 41–48.

Lang, D. (1972). Nucleic acid molecules prepared by monolayer techniques. *Proc. 30th Ann. EMSA Mtg., Los Angeles, Calif.*, pp. 178–79. Claitor's Publishing Div., Baton Rouge, La.

Lee, S. H. (1972). Isolation of parietal cells from glutaraldehyde-fixed rabbit stomach. *J. Histochem. & Cytochem.* **20**, 634.

Leffingwell, H. A., and Hodgkin, N. (1971). Techniques for preparing fossil palynomorphs for study with the scanning and transmission electron microscopes. *Rev. Palaeobotan. Palynol.* **11**, 177.

Lester, D., and Greenberg, L. A. (1950). Acute and chronic toxicity of some halogenated derivatives of methane and ethane. *Arch. Ind. Hygiene Occupat. Med.* **2**, 335.

Lewis, E. (1971). Studying neuronal architecture and organization with the scanning electron microscope. *Proc. 4th Ann. SEM Symp.*, IITRI, Chicago, pp. 281–88.

Lewis, E. R., and Nemanic, M. K. (1973). Critical point drying techniques. In: *Scanning Electron Microscopy/1973* (Johari, O., ed.), pp. 767–74. IITRI, Chicago.

Lim, D. (1971). Scanning electron microscopic observation on non-mechanically cryofractured biological tissue. *Proc. 4th Ann. SEM Symp.*, IITRI, Chicago, pp. 257–64.

Lim, D. J., and Lane, W. C. (1969). The scanning electron microscopic observation of the vestibular sensory epithelia. *Proc. 27th Ann. EMSA Proc.*, pp. 40–41. Claitor's Publishing Div., Baton Rouge, La.

Lung, B. (1968). Whole-mount electron microscopy of chromatin and membranes in bull and human sperm heads. *J. Ultrastruct. Res.* **22**, 485.

MacKenzie, A. P. (1972). Freezing, freeze-drying, and freeze-substitution. *Proc. 5th Ann. SEM Symp.*, IITRI, Chicago, pp. 273–80.

Mackey, B. M., and Morris, J. G. (1972). The exosporium of *Clostridium pasteurianum. J. Gen. Microbiol.* **73**, 325.

Marszalek, D. S., and Small, E. B. (1969). Preparation of soft biological materials for scanning electron microscopy. *Proc. 2nd Ann. SEM Symp.*, IITRI, Chicago, pp. 233–39.

Maser, M. D., Powell, III, T. E., and Philpott, C. W. (1967). Relationships among pH, osmolality, and concentration of fixative solutions. *Stain Tech.* **42**, 175.

Merck Index (1968). 8th ed. (Stecher, P. G., *et al.*, eds.), pp. 1281–98. Merck & Co., Inc., Rahway, N.J.

Mozingo, H. N., Klein, P., Zeevi, Y., and Lewis, E. R. (1970). Venus's flytrap observations by scanning electron microscopy. *Amer. J. Bot.* **57**, 593.

Mueller, J. C., Jones, A. L., and Brandborg, L. L. (1973). Scanning electron microscope observation in human giardiasis. In: *Scanning Electron Microscopy/1973* (Johari, O., ed.), pp. 557–64. IITRI, Chicago.

Nei, T., Yotsumoto, H., Hasegawa, Y., and Nagasawa, Y. (1972). Electron microscopic observation of biological specimens in their native state by employing cryogenic techniques. *Proc. 30th Ann. EMSA Proc.*, pp. 410–11. Claitor's Publishing Div., Baton Rouge, La.

Nemanic, M. K. (1972). Critical point drying, cryofracture, and serial sectioning. *Proc. 5th Ann. SEM Symp.* IITRI, Chicago, Illinois, pp. 297–304.

Nemanic, M. K., and Pitelka, D. R. (1971). A scanning electron microscope study of the lactating mammary gland. *J. Cell Biol.* **48**, 410.

Nowell, J. A., Pangborn, J., and Tyler, W. S. (1972). Stabilization and replication of soft tubular and alveolar systems a scanning electron microscope study of the lung. *Proc. 5th Ann. SEM Symp.*, IITRI, Chicago, pp. 305–12.

Okada, T., and Comings, D. (1970). Whole mount electron microscopy of chicken microchromosomes. *J. Cell Biol.* **47**, 150a.

Otaka, T., and Honjo, S. (1972). A new freeze dry technique for preparation of marine biological specimens for SEM. *Proc. 5th Ann. SEM Symp.*, IITRI, Chicago, pp. 359–363.

Overman, J. (1963). Critical point preparations of intact cells for electron microscopy. *Proc. Soc. Exp. Biol. Med.* **113,** 707.

Pachter, B. R., Penha, D., Davidowitz, J., and Breinin, G. M. (1973). Technique for examining uncoated specimens in the SEM with light microscope and TEM correlation. In: *Scanning Electron Microscopy/1973* (Johari, O., ed.), pp. 387–94. IITRI, Chicago.

Paerl, H. W. (1973). Detritus in Lake Tahoe: Structural modification by attached microflora. *Science* **180,** 496.

Panessa, B. J., and Gennaro, J. F., Jr. (1972a). Preparation of fragile botanical tissues and examination of intracellular contents by SEM. *Proc. 5th Ann. SEM Symp.,* IITRI, Chicago, pp. 327–34.

Panessa, B. J., and Gennaro, J. F., Jr. (1972b). A method for direct observation of botanical tissue and intracellular contents by SEM. *Proc. 30th Ann. EMSA Mtg.,* pp. 208–9. Claitor's Publishing Div., Baton Rouge, La.

Panessa, B. J., and Gennaro, J. F., Jr. (1973). Use of potassium iodide/lead acetate for examining uncoated specimens. In: *Scanning Electron Microscopy/ 1973* (Johari, O., ed.), pp. 395–402. IITRI, Chicago.

Parducz, B. (1967). Ciliary movement and coordination in ciliates. *Int. Rev. Cytol.* **21,** 91.

Pasternak, J. J., Thompson, J. E., Schultz, T. M. G., and Zachariah, K. (1970). A scanning electron microscopic study of the encystment of *Acanthamoeba castellanii. Exp. Cell Res.* **60,** 290.

Paulin, J. J., and Bussey, J. (1971). Oral regeneration in the ciliate *Stentor coeruleus:* A scanning and transmission electron optical study. *J. Protozool.* **18,** 201.

Perecko, J. P., Berezesky, I. K., and Grimley, P. M. (1973). Surface features of some established murine cell lines under various conditions of oncogenic virus infection. In: *Scanning Electron Microscopy/1973* (Johari, O., ed.), pp. 521–28. IITRI, Chicago.

Pfefferkorn, G. E. (1970). Specimen preparation techniques. *Proc. 3rd Ann. SEM Symp.,* IITRI, Chicago, pp. 89–96.

Pfefferkorn, G. E., Gruter, H., and Pfautsch, M. (1972). Observations on the prevention of specimen charging. *Proc. 5th Ann. SEM Symp.,* IITRI, Chicago, pp. 147–52.

Pihl, E., and Bahr, G. F. (1970a). A new approach to the study of cell organelles with the electron microscope. *Exp. Cell Res.* **59,** 379.

Pihl, E., and Bahr, G. F. (1970b). Matrix structure of critical-point dried mitochondria. *Exp. Cell Res.* **63,** 391.

Price, H. L., and Dripps, R. D. (1965). General anesthetics. In: *The Pharmacological Basis of Therapeutics* (Goodman, L. D., and Gilman, A., eds.), 4th ed., pp. 71–73. The Macmillan Company, New York.

Price, Z. (1972). A three-dimensional model of membrane ruffling from transmission and scanning electron microscopy of cultured monkey kidney cells (LLCMK$_2$). *J. Microscopy* **95,** 493.

Rebhun, L. I. (1972). Freeze-substitution and freeze-drying. In: *Principles and Techniques of Electron Microscopy*, Vol. 2 (Hayat, M. A., ed.), pp. 3–49. Van Nostrand Reinhold Company, New York and London.

Richter, I. E., Vogel, K., and Huber, H. J. (1969). Die Untersuchung unfixierter pflanzlicher Objekte mit dem Raster-Elektronenmikroskop "Stereoscan." *Z. f. wiss. Mikroskopie* **69**, 94.

Ris, H., and Kubai, D. F. (1970). Chromosome structure. *Ann. Rev. Gen.* **4**, 263.

Rowe, V. K., and Wolf, M. A. (1963). Ketones. I. General considerations. In: *Industrial Hygiene and Toxicology* (Fassett, D. W., and Irish, D. D., eds.), Vol. II, 2d ed., pp. 1719–70. Wiley Interscience Publishers, New York.

Rowlinson, J. S. (1958). The properties of real gases. In: *Encyclopedia of Physics*, Vol. 12 (Flügge, S., ed.), pp. 37–48, 65–72. Springer-Verlag, Berlin.

Sandborn, E., and Makita, T. (1972). Scanning and transmission electron microscopy of intracellular organelles. *J. Cell Biol.* **55**, 71a.

Satir, P. (1963). Studies on cilia: The fixation of the metachronal wave. *J. Cell Biol.* **18**, 345.

Shimamura, A., and Tokunaga, J. (1970). Scanning electron microscopy of sensory (fungiform) papillae in the frog tongue. *Proc. 3rd Ann. SEM Symp.*, IITRI, Chicago, pp. 225–32.

Small, E., and Marszalek, D. (1969). Scanning electron microscopy of fixed, frozen, and dried protozoa. *Science* **163**, 1064.

Small, E., and Ranganathan, V. S. (1970). The direct study of polluted stream ciliated protozoa via scanning electron microscopy. *Proc. 3rd Ann. SEM Symp.*, IITRI, Chicago, pp. 179–84.

Smith, M. E., and Finke, E. H. (1972). Critical point drying of soft biological material for the scanning electron microscope. *Investigative Ophthalmology* **11**, 127.

Stewart, G. J., Ritchie, W. G. M., and Lynch, P. R. (1973). A scanning and transmission electron microscopic study of canine jugular veins. In: *Scanning Electron Microscopy/1973* (Johari, O., ed.), pp. 473–80. IITRI, Chicago.

Strangeways, T., and Canti, R. (1927). The living cell in vitro as shown by dark ground illumination and the changes induced in such cells by fixing reagents. *Quart. J. Micr. Sci.* **71**, 1.

Sträuli, P., and Haemmerli, G. (1971). Usefulness of instantaneous fixation procedure of Parducz for the demonstration of cell surface specializations in scanning electron microscopy. *J. Microscopy* **95**, 519.

Stubblefield, E., and Wray, W. (1971). Architecture of the Chinese hamster metaphase chromosome. *Chromosoma* **32**, 262.

Swift, J. A., and Brown, A. C. (1970). An environmental cell for the examination of wet biological specimens at atmospheric pressure by transmission scanning electron microscopy. *J. Physics E: Sci. Instr.* **3**, 924.

Sybers, H. D., and Ashraf, M. (1973). Preparation of cardiac muscle for SEM. In: *Scanning Electron Microscopy/1973* (Johari, O., ed.), pp. 341–48. IITRI, Chicago.

Tanaka, K., and Iino, A. (1972). Frozen resin cracking method for scanning electron microscopy and its application to cytology. *Proc. 30th Ann. EMSA Mtg.*, pp. 408–9. Claitor's Publishing Div., Baton Rouge, La.

Tanaka, T., Kosaka, N., Takiguchi, T., Aoki, T., and Takahara, S. (1973). Observation on the cochlea with SEM. In: *Scanning Electron Microscopy/1973* (Johari, O., ed.), pp. 427–34. IITRI, Chicago.

Tegner, M. J., and Epel, D. (1973). Sea urchin sperm-egg interactions studied with the scanning electron microscope. *Science* **179**, 685.

Treon, J. F. (1963). Alcohols. In: *Industrial Hygiene and Toxicology*, Vol. II (Fassett, D. W., and Irish, D. D., eds.), 2d ed., pp. 1409–96. Wiley Interscience Publishers, New York.

Turnbill, C., and Philpott, D. E. (1970). The critical point drying method applied to scanning electron microscopy of L-929 cells. *Proc. 28th Ann. EMSA Mtg.*, pp. 278–79. Claitor's Publishing Div., Baton Rouge, La.

Vesely, P., and Boyde, A. (1973). The significance of SEM evaluation of the cell surface for tumor cell biology. In: *Scanning Electron Microscopy/1973* (Johari, O., ed.), pp. 689–96. IITRI, Chicago.

Westfall, J. A., and Enos, P. D. (1972). Scanning and transmission electron microscopy of isolated cells of *Hydra littoralis*. *Proc. 30th Ann. EMSA Mtg.*, pp. 160–61. Claitor's Publishing Div., Baton Rouge, La.

Wetzel, B., Erickson, B. W., Jr., and Levis, W. R. (1973). The need for positive identification of leukocytes examined by SEM. In: *Scanning Electron Microscopy/1973* (Johari, O., ed.), pp. 535–42. IITRI, Chicago.

Wickham, M. G., and Worthen, D. M. (1973). Correlation of scanning and transmission electron microscopy on the same tissue sample. *Stain Tech.* **48**, 63.

Widom, B. (1967). Intermolecular forces and the nature of the liquid state. *Science* **157**, 375.

Williams, A. E., Jordan, J. A., Murphy, J. F., and Allen, J. M. (1973). The surface ultrastructure of normal and abnormal cervical epithelia. In: *Scanning Electron Microscopy/1973* (Johari, O., ed.), pp. 597–604. IITRI, Chicago.

Wohlfarth-Bottermann, K. E., and Komnick, H. (1966). Die Gefahren der Glutaraldehyd-Fixation. *J. Microscopie* **5**, 441.

Wolfe, S. L. (1965). Isolated microtubules. *J. Cell Biol.* **25**, 408.

Wolfe, S. L., and Grim, J. N. (1967). The relationship of isolated chromosome fibers to the fibers of the embedded nucleus. *J. Ultrastruct. Res.* **19**, 382.

Wolfe, S. L., and Martin, P. G. (1968). The ultrastructure and strandedness of chromosomes from two species of *Vicia*. *Exp. Cell Res.* **50**, 140.

Wollman, H., and Dripps, R. D. (1965). The therapeutic gases oxygen, carbon dioxide, and helium. In: *The Pharmacological Basis of Therapeutics* (Goodman, L. S., and Gilman, A., eds.), 4th ed., 908–29. The Macmillan Company, New York.

Worthen, D. M., and Wickham, M. G. (1972). Scanning electron microscopy tissue preparation. *Investigative Ophthalmology* **11**, 133.

Zirkin, B. (1971). The fine structure of nuclei in mature sperm. I. Application of the Langmuir trough-critical point method to histone-containing sperm nuclei. *J. Ultrastruct. Res.* **36**, 237.

3. CRYOTECHNIQUES

Tokio Nei

The Institute of Low Temperature Science
Hokkaido University, Sapporo, Japan

INTRODUCTION

Most biological materials contain 70 to 80% water. In order to prepare biological specimens for electron microscopy, it is necessary to dehydrate them. However, the removal of water may cause alterations in the specimens. In conventional preparation, the specimens must be subjected to physical and chemical treatments such as fixation, dehydration, embedding, sectioning, and staining. These treatments can cause artifacts (Hayat, 1970).

It has long been hoped to observe the specimens in their native state with an electron microscope. To accomplish this, considerable efforts have been made to develop suitable techniques. Two of the approaches are: the chamber or capsule method (Nagata and Ishikawa, 1972), and cryotechniques (Bullivant, 1970; MacKenzie, 1972). The former is still to be perfected, and has not yielded successful results as yet. The latter, on the other hand, has been developed extensively.

For morphological studies with an electron microscope, freeze-drying and freeze-substitution have been most commonly employed to avoid the artifacts produced during dehydration with solvents (Rebhun, 1972). Nevertheless, it is still difficult to completely prevent the shrinkage or deformation of the specimen during freeze-drying and to reduce the release or loss of chemical substances during freeze-substitution. The freeze-etching technique, on the other hand, has an advantage over the other two techniques in that there is very little or no dehydration involved. The limitation of freeze-etching is that only a replica of the specimen can be observed under the transmission electron microscope (Koehler, 1972).

Direct observation of frozen specimens in the transmission electron microscope has been attempted by Nei (1970). The results obtained in these studies using microbial suspension showed that the internal fine structure of the hydrated cells was difficult to visualize, owing to cell density, unless the cells were subjected to some dehydration (Nei, 1962a).

In order to observe the natural configuration of biological specimens, further efforts were then directed toward the scanning electron microscopy of frozen specimens. It has been shown that hydrated specimens in the frozen state can be observed directly with a scanning electron microscope. Echlin *et al.* (1970) first reported successful results using several kinds of plant specimens. Nei *et al.* (1971b, 1972, and 1973) also attempted to observe frozen specimens of plants and animals; particular attention was given to the internal structure, which was exposed by fracturing, as well as to the natural surface.

AIMS OF CRYOTECHNIQUES

The first objective of cryotechniques is to maintain the three-dimensional structure of the specimen. Although most biological specimens are likely to shrink or deform when subjected to air-drying, their original three-dimensional structure can be retained by freeze-drying. The second objective is to solidify the soft tissues by freezing. Solidification makes the tissue easy to be sectioned or fractured. The third objective is to lessen the release of chemical substances. Chemical treatments, especially dehydration with organic solvents employed in conventional processing, may cause the release of some chemical constituents from cells and tissues. Dehydration of specimens by sublimation from the frozen state inhibits such losses. The fourth objective is to observe the specimen in its native state. If the specimen is kept at sufficiently low temperatures in the microscope, there is no dehydration (sublimation) even under high vacuum. It should therefore be possible to observe the frozen specimen directly with the scanning electron microscope. The fifth objective is to maintain the viability of the specimen. The activity or viability of biological specimens can be maintained in the frozen state at low temperatures. With this approach, both morphological and physiological studies can be carried out concurrently.

FACTORS AFFECTING THE MORPHOLOGY OF SPECIMENS IN CRYOTECHNIQUES

Freezing patterns of biological specimens depend largely upon the rate of cooling and the concentration of antifreeze agents. The most serious artifact that might arise in the specimen prepared according to cryo-techniques is ice crystal formation. When a biological specimen is cooled below its freezing point, water turns into ice crystals which grow larger with further drop in temperature. The faster the cooling, the smaller the ice crystals will be formed. The reason for this is that a high rate of cooling increases the rate of nucleation, and hence a reduction in the crystal size. It is desirable that ice crystals formed during freezing should be as small as possible. The formation of ice crystals larger than the resolution limit of the electron microscope used should be avoided; otherwise the images obtained would be incorrectly interpreted.

The ideal cryotechnique is to attain a vitreous (amorphous or glassy) state by controlling the rate of cooling and the concentration of anti-freeze agents. If this is accomplished, the natural configuration of hydrated specimens can be observed, even with an electron microscope. For achieving this goal, the following factors should be considered.

Rate of Cooling

Specimen Size. The total mass of the specimen should be reduced as much as possible in order to achieve a high rate of cooling. However, there is some limitation in minimizing the size that can be handled.

Coolant. The ideal coolant is characterized by having a high boiling point and a low freezing point as well as a high thermal conductivity and heat capacity. Suitable coolants include propane, isopentane, Freon 12, and Freon 22. Liquid nitrogen is not an ideal direct coolant, because evaporated nitrogen gas insulates the specimen and reduces the rate of cooling (Pease, 1967).

Inhibition of Ice Crystal Formation

Partial Dehydration. Partial dehydration of specimens is sometimes effective for lowering the freezing point and reducing the formation of ice crystals.

Antifreeze Agents. Cryoprotectants such as chloroform, glycerol, and DMSO, which were originally used as additives for protecting viable organisms from freezing injury (Lovelock, 1953a and b), are also useful for reducing the ultimate size of ice crystals. This is believed to be either due to the inhibition of aggregation of water molecules that form ice crystals or to the reduction in the total amount of water available to form ice crystals (Boyde and Wood, 1969). Chloroform has the advantage of not leaving residues on cryofractured tissues.

The relationship between the rate of cooling and the concentration of glycerol has been examined (Nei *et al.*, 1971a). These experiments using human erythrocytes showed that the addition of 10 to 20% glycerol was insufficient to prevent the formation of ice crystals during rapid freezing. Specimens to which 30% glycerol had been added showed no ice formation, but recrystallization of ice occurred at relatively low temperatures (e.g., $-80°C$) during the rewarming process of rapidly frozen specimens (Nei, 1971 and 1973; Nei and Asada, 1972).

Because concentrated glycerol is usually toxic to living cells, it has normally been used as a protective additive in concentrations of 10 to 15%. In cryotechniques also, high concentrations of glycerol might be injurious to living cells.

The use of glycerol or DMSO in preparing frozen tissues for scanning electron microscopy is discouraged, because of the low vapor pressure of these coolants which are retained in the sample. Boyde and Wood (1969) recommended the use of freeze-drying following substitution of water with organic liquids. Specimens that have been frozen and freeze-dried after the substitution of their water with pure, nonpolar organic solvents (e.g., amyl acetate) appear to show few artifacts due to ice crystal growth. Freeze-drying of specimens frozen in nonpolar solvents has the advantage of being very rapid because these solvents have a high vapor pressure below their freezing points. Freeze-drying in the presence of amyl acetate is completed in only half an hour at $-75°C$ (melting point of amyl acetate is $-71°C$) under a vacuum of 5×10^{-3} torr, whereas freeze-drying in the presence of water requires up to a week at $-70°C$. The disadvantages in solvent substitution are slight tissue shrinkage and chemical extraction.

The use of a volatile reagent in place of glycerol was studied by Haggis (1972). He attempted prefixation with 2% glutaraldehyde in sucrose buffer and impregnation in a mixture of dioxane (49.5%), water (49.5%), and glycerol (1%). This was followed by rapid freezing, fracturing, and vacuum drying at $-80°C$. Ice crystal formation was minimized by using this procedure.

PRACTICAL APPLICATIONS

Freeze-Drying

Since its first application in 1890 by Altmann, freeze-drying has been employed for dehydration to prepare the specimens for morphological studies with the microscope. The primary aims of this technique include the maintenance of three-dimensional structure of the specimen during dehydration and the completion of dehydration without releasing chemical constituents of the specimen. During conventional preparatory procedures, air-drying causes deformation of the specimens, and chemical dessicants (organic solvents) cause release of chemical constituents of the specimen. Although freeze-drying is superior to air-drying, it is difficult to avoid completely the shrinkage that occurs in the final stage of drying (Nei, 1962b).

The specimen is rapidly frozen by immersing in Freon 22 that has been cooled to $\sim -150°C$ with liquid nitrogen. It is then transferred to a precooled container which is connected to a freeze-dryer. Sublimation in the frozen specimen begins with evacuation. The temperature of the specimen should be kept as low as possible (at least below $-70°C$), even though this will prolong the duration of drying. The duration of drying depends primarily upon the specimen temperature and size. After drying, the specimen is placed in an evacuation unit and coated with gold while revolving on a rotatory stage. The specimen is now ready for observation with a scanning electron microscope.

Freeze-Drying with Cryofracture

The cryofracture technique has two advantages. First, the fracture follows the plane of structural weakness, and may sometimes take a direction through the tissue revealing more than that seen when it is sectioned in one plane. Second, cryofracture takes place at a much lower temperature than those used for cutting frozen sections. Also, it almost eliminates the attendant ice crystal growth and recrystallization (Haggis, 1970).

The specimen is fixed with glutaraldehyde and washed in distilled water. It is then rapidly frozen in Freon 22 at $-150°C$, fractured at $-170°C$ with a cold blade, and freeze-dried under vacuum at $-80°C$ for 15 hr. Alternatively, the specimen can be frozen directly without prior fixation or washing. Washings aid in the removal of soluble materials from the tissue; otherwise they may obscure the morphology in the electron micrograph. In the specimens thus prepared, rather large crevices often open up between the cells, thereby exposing their outer surfaces. Ice crystals of various dimensions, depending upon the characteristics of the specimen, are found in cells or solutions (Haggis, 1970).

Nonmechanical cryofracture technique has been employed for preparing delicate tissues (e.g., inner ear) (Lim, 1971). This method minimizes the artifacts caused by shearing forces. Specimens are fixed with buffered 1% osmium tetroxide or with buffered 2% glutaraldehyde. The specimens are washed in saline, cut into small pieces (0.1 x 2 mm), and dehydrated with ethanol up to 70% solution. The tissue suspension in 70% ethanol is dropped with a pipette into an aluminum dish dipped in liquid nitrogen. The droplet instantly freezes and sinks to the bottom, where it eventually cracks. These frozen droplets are transferred to a tissue dryer and freeze-dried at −80°C for 5 to 24 hr. Optimum results were obtained by using 70% ethanol at room temperature and a drop size of ∼4 mm.

The mechanism of automatic fracturing described above is not clear. One possible explanation is that sudden expansion of the gas trapped inside the tissue or ethanol droplet may cause the fracturing, or that sudden thermal expansion is responsible for this phenomenon.

Direct Observation of Frozen Specimens

Attempts have been made to examine frozen specimens without dehydration and metal-coating directly under the scanning electron microscope (Echlin et al., 1970; Nei et al., 1971b). A specimen stub of 13 mm diameter is screwed into a support block. Tilts of up to 50° and rotation of ∼45° can be applied. The temperature control of the stage is obtained by a system analagous to a feedback system.

Specimens are mounted on the specimen stub and immediately plunged into Freon 22 maintained at its melting point (−160°C) by liquid nitrogen. The specimen stub is rapidly transferred to the cold stage maintained at −180°C with liquid nitrogen, and the microscope column is pumped down to 1 x 10^{-4} torr.

Freon 22 quench-freezing yields a specimen almost free of ice crystal damage. It is not advantageous to impregnate the specimens with glycerol or DMSO prior to rapid freezing. The antistatic spray, Duron, also yields poor results because it obscures surface details. The best images are obtained by fixation with 2% glutaraldehyde for 2 hr and postfixation with 1% osmium tetroxide for 30 min; both the fixatives are prepared in phosphate buffer (Echlin, 1971).

The examination of the specimen revealed the surface covered with ice which is formed during the specimen transfer from the quenching coolant to the specimen stage. The source of this ice is atmospheric water that undergoes condensation. The temperature of the stage is raised to between −100 and −90°C, whereupon the high vacuum within the microscope column causes the surface water to sublime, revealing the tissue surface below. The removal of water must be carefully monitored, and the temperature should not be allowed to rise above −85°C. If too much water is removed, by allowing too great a rise in temperature, the tissue will collapse.

Direct Observation of Frozen Specimens with Cryofracture

Cryofracturing of frozen specimens in the scanning electron microscope was attempted by Nei *et al.* (1971b and 1972). These investigators used two types of cold stages incorporated in the microscope. One cold stage is installed in the specimen chamber of the microscope to maintain the specimen holder at temperatures of $\sim -150°C$ during most observations. The second cold stage, designed to maintain the frozen state of the specimen and concurrently to cool a knife for cryofracturing at $-100°C$ or below, is incorporated in the pre-evacuation chamber.

A small piece (2 mm^3) of the tissue is mounted on the specimen stub and held in position with a tiny sharp hook or needle. The specimen holder is quickly immersed into Freon 22 cooled with liquid nitrogen. After this rapid freezing, the holder is quickly transferred to the cold stage of the specimen chamber. In order to prevent moisture condensation during this transfer, the specimen is covered with a metal plate that has previously been installed on the holder. When the metal is not used, the frost which condenses on the surface of the frozen specimen can be sublimated by a temperature rise to $\sim -90°C$ under high vacuum. The sublimation of ice from the frozen specimen is also employed prior to observing cells suspended in a liquid medium. The specimen temperature should be maintained at sufficiently low temperatures to inhibit further dehydration from the frozen specimen surface; otherwise the electric charge at the dehydrated part will disturb the image under scanning, and shrinkage of the tissue may occur.

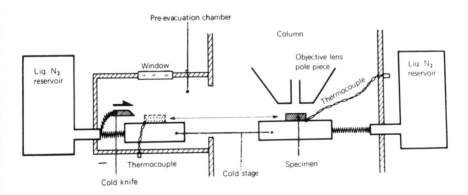

Fig. 3–1. Schematic illustration of cryounit processes.

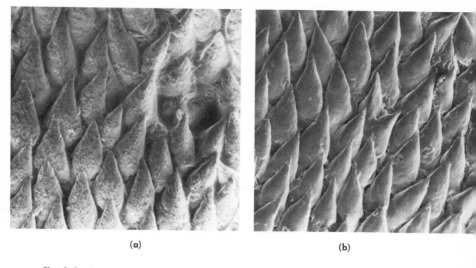

(a)   (b)

Fig. 3–2. Hamster tongue. Frozen specimen surface before and after defrosting. The specimens were neither dehydrated nor coated. 65 X.

(a)   (b)

Fig. 3–3. Cross sections of hamster tongue, cryofractured and uncoated. (a) papilla; (b) muscle fibers in the deep layer. 200 X.

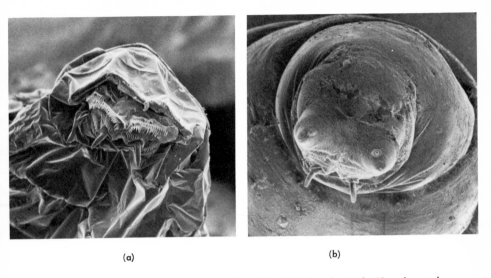

(a)   (b)

Fig. 3–4. Fruit fly (*Drosophila meganogaster*) larva. (a) air-dried and coated with carbon and gold; (b) frozen and uncoated. Note the remarkable difference in the appearance of the two preparations. 65 X.

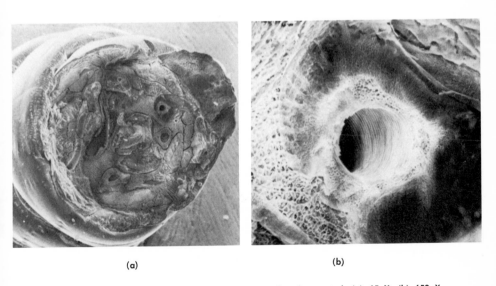

(a)   (b)

Fig. 3–5. Cross sections of fruit fly larva, cryofractured and uncoated. (a) 65 X; (b) 650 X.

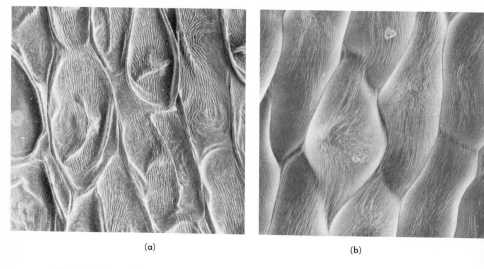

(a)   (b)

Fig. 3–6. Chrysanthemum petals. (a) air-dried and coated; (b) frozen and uncoated. 300 X.

The specimen thus prepared is first observed at the surface from various angles under the scanning electron microscope with the accelerating voltage of 5kV. This kV is preferred over a high accelerating voltage, because the latter will cause charging of the uncoated specimen. Lower voltages, however, will result in some decrease in resolution. The specimen is then replaced on the cold stage in the pre-evacuation chamber and fractured at a desired portion with a cold knife at temperatures below −100°C. The fractured face is observed after the specimen is transferred to the cold stage in the specimen chamber (refer to figures).

CONCLUSIONS

It is assumed that frozen specimens of hydrated specimens without dehydration and metal-coating reveal their original three-dimensional structures. Cryofracture yields an appearance quite different from the collapsed shape caused by air-drying. The former is a simple and useful technique for observing the internal structure of soft tissues in their native state. However, the size of a specimen frozen at a high rate of cooling is determined by the temperature gradient which must occur between the peripheral and internal parts. This sometimes causes rather large ice crystals in the core of the specimen, which becomes the most serious problem in the observation of fractured faces. The use of anti-freeze agents such as glycerol is not desirable because they obscure the details of the fine structure. It is known that some of these agents are retained by the specimen.

Although many problems still remain to be solved, cryotechniques provide the opportunity for directly visualizing the native configuration of biological specimens with a scanning electron microscope. Further development and application of this technique are expected in the near future.

REFERENCES

Altmann, R. (1890). *Die Elementarorganismen und ihre Beziehungen zur den Zellen.* Leipzig.

Boyde, A., and Wood, C. (1969). Preparation of animal tissues for surface-scanning electron microscopy. *J. Microscopy* 90, 221.

Bullivant, S. (1970). Present status of freezing techniques. In: *Some Biological Techniques in Electron Microscopy* (Parsons, D. F., ed.), Academic Press, New York. p. 101.

Echlin, P. (1971). The examination of biological material at low temperatures. *Scanning Electron Microscopy. Proc. 4th Ann. Scan. E. M. Symp.,* I.I.T. Research Institute, Chicago, p. 225.

Echlin, P., Paden, R., Dronzek, B., and Wayte, R. (1970). Scanning electron microscopy of labile biological material maintained under controlled conditions. *Scanning Electron Microscopy. Proc. 3rd Ann. Scan. E. M. Symp.,* I.I.T. Research Institute, Chicago, p. 49.

Haggis, G. H. (1970). Cryofracture of biological material. *Scanning Electron Microscopy. Proc. 3rd Ann. Scan. E. M. Symp.,* I.I.T. Research Institute, Chicago, p. 97.

Haggis, G. H. (1972). Freeze-fracture for scanning electron microscopy. *Proc. 5th European Congr. Electron Micros.,* p. 250.

Hayat, M. A. (1970). *Principles and Techniques of Electron Microscopy: Biological Applications,* Vol. 1. Van Nostrand Reinhold Company, New York and London.

Koehler, J. K. (1972). The freeze-etching technique. In: *Principles and Techniques of Electron Microscopy: Biological Applications,* Vol. 2 (Hayat, M. A., ed.). Van Nostrand Reinhold Company, New York and London.

Lim, D. J. (1971). Scanning electron microscopic observations on non-mechanically cryofractured biological tissue. *Scanning Electron Microscopy. Proc. 4th Ann. Scan. E. M. Symp.,* I.I.T. Research Institute, Chicago, p. 257.

Lovelock, J. E. (1953a). The haemolysis of human red blood-cells by freezing and thawing. *Biochim. Biophys. Acta* 10, 414.

Lovelock, J. E. (1953b). The mechanism of the protective action of glycerol against haemolysis by freezing and thawing. *Biochim. Biophys. Acta* 11, 28.

MacKenzie, A. P. (1972). Freezing, freeze-drying, and freeze-substitution. *Scanning Electron Microscopy. Proc. 5th Ann. Scan. E. M. Symp.*, I.I.T. Research Institute, Chicago, p. 273.

Nagata, F., and Ishikawa, I. (1972). Observation of wet biological materials in a high voltage electron microscope. *Jap. J. Appl. Phys.* **11**, 1239.

Nei, T. (1962a). Electron microscopic study of microorganisms subjected to freezing and drying: Cinematographic observations of yeast and coli cells. *Exptl. Cell Res.* **28**, 560.

Nei, T. (1962b). Freeze-drying in the electron microscopy of microorganisms. *J. Electron Micros.* **11**, 51.

Nei, T. (1970). Studies on the freezing and drying of biological specimens by electron microscopy. *J. Electron Micros.* **19**, 6.

Nei, T. (1971). Hemolysis during the rewarming process of frozen erythrocytes. *Proc. XIIIth Intern. Congr. Refrig.* **3**, 907.

Nei, T. (1973). Growth of ice crystals in frozen specimens. *J. Microscopy* (in press).

Nei, T., and Asada, M. (1972). Changes appearing in the rewarming process of rapidly frozen erythrocytes. *Low Temp. Sci.*, B, **30**, 45.

Nei, T., Matsusaka, T., and Asada, M. (1971a). Investigations on the cooling conditions in the freeze-etching technique. *Low Temp. Sci.*, B, **29**, 91.

Nei, T., Yotsumoto, H., Hasegawa, Y., and Nagasawa, Y. (1971b). Direct observation of frozen specimens with a scanning electron microscope. *J. Electron Micros.* **20**, 202.

Nei, T., Yotsumoto, H., Hasegawa, Y., and Nagasawa, Y. (1972). Electron microscopic observations of biological specimens in their native state by employing cryogenic techniques. *Proc. 5th European Congr. Electron Micros.*, p. 252.

Nei, T., Yotsumoto, H., Hasegawa, Y., and Nagasawa, Y. (1973). Direct observation of frozen specimens with a scanning electron microscope. *J. Electron Micros.* **22**, 185.

Pease, D. C. (1967). The preservation of tissue fine structure during rapid freezing. *J. Ultrastruct. Res.* **21**, 98.

Rebhun, L. I. (1972). Freeze-substitution and freeze-drying. In: *Principles and Techniques of Electron Microscopy: Biological Applications*, Vol. 2 (Hayat, M. A., ed.). Van Nostrand Reinhold Company, New York and London.

4. FROZEN RESIN CRACKING METHOD AND ITS ROLE IN CYTOLOGY

Keiichi Tanaka

Department of Anatomy, Tottori University School of Medicine,
Yonago, Japan

INTRODUCTION

Until recently the scanning electron microscope has been employed primarily for studying the surfaces of cells and tissues rather than intracellular structures. The main reason for this has been the nonavailability of satisfactory techniques. A few reports, however, have been published which have revealed intracellular structures. Several recent methods have used: tissue sectioner (Makita and Sandborn, 1971), freeze-fracturing (Boyde and Wood, 1969; Germinario and McAlear, 1971; Humphreys and Wodzicki, 1972), cryofracturing (Haggis, 1970; Lim, 1971; Nemanic, 1972), and aqueous iodine solution (Panessa and Gennaro, 1972). The frozen resin cracking method (Fig. 4–1) (Tanaka, 1972a; Tanaka and Iino, 1972) has yielded satisfactory results, and will be described below along with its cytological application.

Resin Cracking Method for SEM

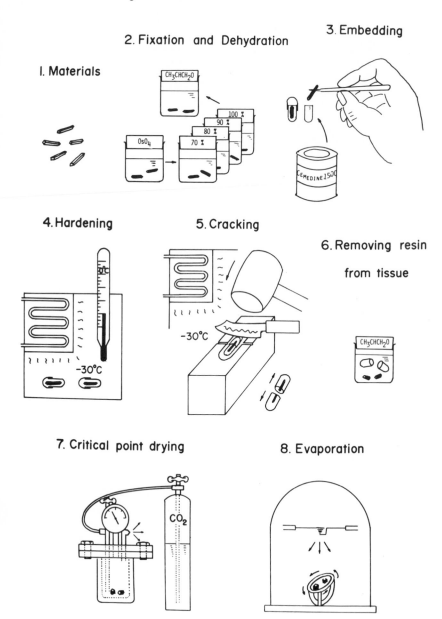

Fig. 4–1. Schematic drawing of frozen resin cracking method.

METHOD

Fixation and Dehydration

Fresh specimens ($1 \times 1 \times 5$ mm) are fixed with glutaraldehyde and osmium tetroxide. The specimens are dehydrated with a graded series of ethanol (50 to 100%) according to standard procedures.

Embedding

The dehydrated specimens are transferred to propylene oxide for 30 min and then placed in a mixture (1:1) of propylene oxide and Cemedine 1500 or Araldite GY260 for 2 to 4 hr; a thorough permeation of the tissue should be assured. The specimens are embedded in small gelatin capsules. The resin is used without adding any catalyst and is warmed before pouring into the capsules. The hardening of the resin is carried out in either a cryostat or a refrigerator at $-30°C$ for 1 to 2 hr (Cemedine or Araldite GY260) or at $-80°C$ (Epon 812 or polyester). The specimen should be moved to the center of the capsule prior to hardening.

Cracking

The capsules are cracked into two pieces in the cryostat with cutlery and a hammer on a cutting board. Epon capsules should be dipped in ether containing dry ice for 3 to 5 min to be hardened and cracked into two pieces at room temperature. The capsules should not be cut but cracked. The cracked surface of the capsule should be smooth and shiny. The cutting board can be made of wood with a hole for holding the capsule (Fig. 4–1). A razor blade, chisel, or any other piece of cutlery can be used as a cracking tool. A thick blade is adequate for obtaining specimens of good quality, which reveal rough external features of the cell (Fig. 4–2).

The resin can be removed from the tissue by placing the cracked capsules into propylene oxide. This treatment is repeated several times at 30 min intervals. When the room temperature is low, the solvent should be warmed slightly.

Drying

The specimen is dried by the critical point drying method (Anderson, 1951 and 1956; Cohen et al., 1968; Boyde and Wood, 1969; Tanaka, 1972b; Cohen, in this volume). Prior to transferring to the critical point drying

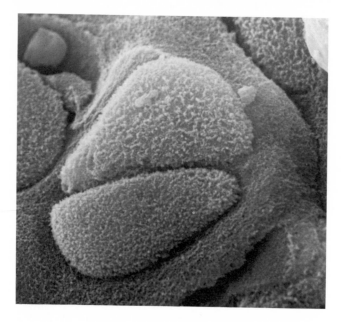

Fig. 4–2. Cartilage cells of a rabbit rib; cell surface is clearly seen. 3,500 X.

apparatus, the specimen must be dipped in amyl acetate for 20 min. Acetone air-drying can be used, but the results are less satisfactory.

Metal-Coating

After the specimen has been dried, the cracked surface can be identified under a stereo-light microscope because it is shiny like the surface of broken coal. After it has been trimmed properly, the specimen is attached to the holder with the shiny side up and coated with evaporated chromium and gold-palladium in a rotary vacuum evaporator.

Since 10,000 to 30,000 times magnification is usually used for observing intracellular structures, the microscope should be maintained at top condition. The high-resolution field emission SEM is also useful for observing the specimen (Fig. 4–3).

The frozen resin cracking method can be practiced easily; a skilled hand is not essential. There seems to be no discernible artifacts. Ultrathin sections can be obtained from cracked and dried specimens for transmission electron microscopy (Fig. 4–4).

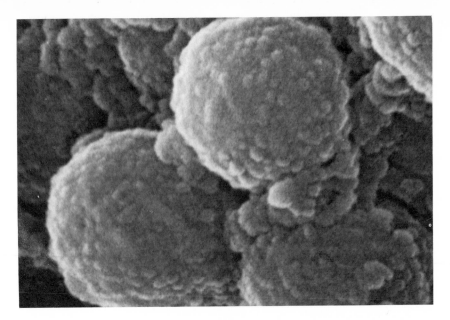

Fig. 4–3. Zymogen granules observed with high-resolution field emission scanning electron microscope. 60,000 X.

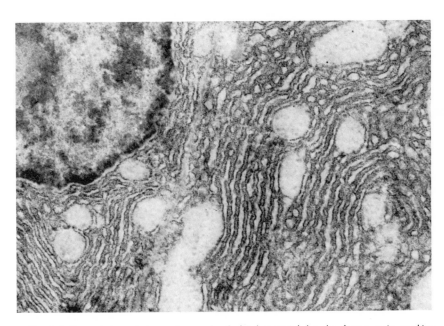

Fig. 4–4. Transmission electron micrograph of dried material by the frozen resin cracking method; pancreatic acinar cell from a dog.

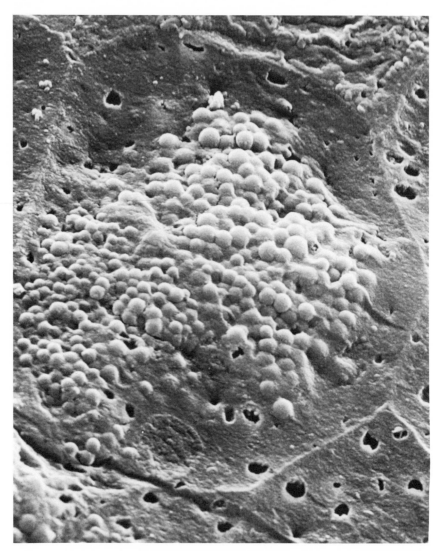

Fig. 4–5. Zymogen granules of pancreatic acinar cells from a dog; a nucleus is also present. 6,000 X.

INTRACELLULAR STRUCTURES

Nucleus and Nucleolus

The nucleus generally appears as an assembly of rough granules surrounded by a nuclear envelope (Fig. 4–5). The nucleolus is seen as a dense clump in the nucleus. If the specimen is treated with hypotonic solutions before fixation, the nucleus shows a netlike structure composed of columns of 2,000 to 3,000 Å in diameter (Fig. 4–6).

Endoplasmic Reticulum

The cisternal and tubular endoplasmic reticulum is seen to be arranged three-dimensionally and studded with small particles of uniform size (50 to 60 nm), which seem to be related to ribonucleoprotein (Fig. 4–7). However, it is not clear whether each particle contains only one ribosome or more than one ribosome (polysome).

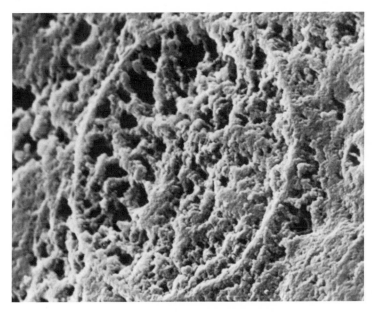

Fig. 4–6. Nucleus from a rabbit liver cell treated with hypnotonic solution; bumpy, netlike structure can be seen. 21,300 X.

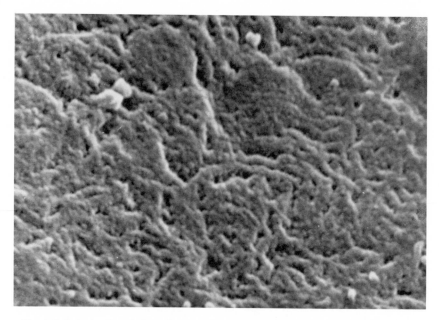

Fig. 4–7. Endoplasmic reticulum of a pancreatic acinar cell from a dog. 42,000 X.

Golgi Complex

The appearance of the Golgi complex is different from that seen with the transmission electron microscope. The Golgi appears to be a collection of match stick heads, which may be the central core of the Golgi complex (Fig. 4–8); the outer lamellar part is not clear.

Mitochondria

Mitochondria are easily recognizable, but the cristae cannot be seen clearly. The reason for this may be that the density of the cristae is almost the same as that of the matrix.

Secretory Granules

Secretory granules are easily visible (Fig. 4–5). When magnified more than 40,000 times by means of a field emission scanning electron microscope, small particles can be seen attached to the granules (Fig. 4–3).

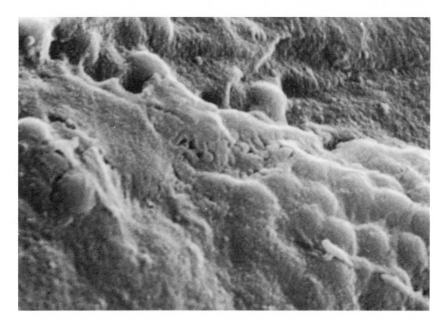

Fig. 4–8. Golgi complex of a pancreatic acinar cell from a dog. 19,000 X.

Other Structures

The fine structure of intercellular substance of cartilage can be revealed by using the above method; collagen fibers and hydroxyl apatite clusters, for instance, have been observed. The surface of small canals in the tissue, such as cilia of an ependymal cell of spinal cord has also been observed (Ohtsuki, 1972). Centrioles have not been identified as yet.

REFERENCES

Anderson, T. F. (1951). Techniques for the preservation of three-dimensional structure in preparing specimens for the electron microscope. *Trans. N. Y. Acad. Sci.*, Ser. II, **13**, 130.

Anderson, T. F. (1956). Electron microscopy of microorganisms. In: *Physical Techniques In Biological Research*, Vol. 3 (Pollister, A. W., ed.), p. 319. Academic Press, New York.

Boyde, A., and Wood, C. (1969). Preparation of animal tissues for surface-scanning electron microscopy. *J. Microscopy* **90**, 221.

Cohen, A. L., Marlow, D. P., and Garner, G. E. (1968). A rapid critical point method of using fluorocarbons ("Freons") as intermediate and transitional fluids. *J. Mikroscopie* **7**, 331.

Germinario, L. T., and McAlear, J. H. (1971). Preparation of tissue for scanning electron microscopy: Freeze-fracturing as a technique for enhancing visibility of structural relationships. *Stain Technol.* **46**, 249.

Haggis, C. H. (1970). Cryofracture of biological material. In: *Scanning Electron Microscopy.* I.I.T. Research Institute, Chicago, p. 99.

Humphreys, W. J., and Wodzicki, T. J. (1972). Methods for viewing by electron microscopy the interior organization of protoplasts of plant cells. *Proc. 30th Ann. Meet. EMSA,* p. 238.

Lim, D. J. (1971). Scanning electron microscopic observation on non-mechanically cryofractured biological tissue. In: *Scanning Electron Microscopy.* I.I.T. Research Institute, Chicago, p. 257.

Makita, T., and Sandborn, E. B. (1971). Identification of intracellular components by scanning electron microscopy. *Exptl. Cell Res.* **67**, 211.

Nemanic, M. K. (1972). Critical point drying, cryofracture and serial sections. In: *Scanning Electron Microscopy.* I.I.T. Research Institute, Chicago, p. 297.

Ohtsuki, K. (1972). Scanning electron microscopic studies on rabbit's spinal cord by resin cracking method. *Arch. Histol. Jap.* **34**, 405.

Panessa, B. J., and Gennaro, J. F., Jr. (1972). A method for direct observation of botanical tissue and intracellular contents by SEM. *Proc. 30th Ann. Meet. EMSA,* p. 208.

Tanaka, K. (1972a). Frozen resin cracking method for scanning electron microscopy of biological materials. *Naturwiss.* **59**, 77.

Tanaka, K. (1972b). A simple type of apparatus for critical point drying method. *J. Electron Microscopy* **21**, 153.

Tanaka, K., and Iino, A. (1972). Frozen resin cracking method for scanning electron microscopy and its application to cytology. *Proc. 30th Ann. Meet. EMSA,* p. 408.

5. PREPARATION OF STEREO SLIDES FROM ELECTRON MICROGRAPH STEREOPAIRS

Michael K. Nemanic

Department of Zoology and Its Cancer Research Laboratory,
University of California, Berkeley, California

INTRODUCTION

Research using scanning electron microscopy (Boyde, 1970; Nemanic, 1972b), transmission electron microscopy of freeze-etch replicas (Staehelin, 1970), or high-voltage transmission electron microscopy of thick sections (Hama and Nagata, 1970) often needs the aid of recorded three-dimensional images of the specimen under study. The addition of a third dimension to the two-dimensional images normally seen is accomplished in publication by printing paired stereomicrographs. Projection of stereo images on a viewing screen is more difficult to achieve, since they must be aligned so that when viewed through appropriate glasses they yield a stereoscopic image. The methods used for the preparation and projection of stereo slides are the subject of this chapter. Stereo measurements and interpretation of stereo images will not be discussed. Articles by Boyde (1970) and Howell and Boyde (1972) are suggested readings in these two areas.

PREPARATION OF STEREOMICROGRAPHS

A stereopair is composed of two micrographs which show the same specimen area viewed at an angular displacement that usually conforms to normal human stereoscopic vision (i.e., ~7°). The tilt angle for stereopairs of transmission sections is variable, however, depending on the specimen thickness and the final magnification used to view the pair (Thomas and Lentz, 1972).

The procedure for taking the two exposures of a stereopair is simple. After the first micrograph of the pair is taken, the sample is tilted with respect to the electron beam in the microscope column. In most electron microscopes, the sample is tilted by manipulating a tilting stage; however, some high-voltage transmission microscopes tilt the electron beam rather than the sample. In the transmission microscope, the tilted image is refocused by varying the current in the objective lens and recentered with the stage controls. Some transmission stages tilt the specimen about the axis of the electron beam, minimizing the need to refocus and recenter. In the scanning electron microscope, the image is refocused with the z-axis control and recentered exactly by the x- and y-axis controls (Boyde, 1971a). After the focusing and centering procedures are completed, the second micrograph of the pair is taken.

In all stereo slides, the tilt axis must be aligned vertically; stereopairs taken on a transmission microscope can have an arbitrary tilt direction (i.e., a tilt axis not parallel with one of the edges of a photographic print) (Thomas and Lentz, 1972). Before a stereo slide can be made, this tilt direction must be determined by rotating the individual members of a stereopair until the best stereo impression is achieved. Stereopairs taken on a scanning microscope generally have a fixed tilt axis, because most stages have only one tilt direction, usually parallel with the edge of a micrograph.

Several methods are used for preparing slides from electron micrograph stereopairs. The object of these procedures is to project and view both members of the pair in such a way that the right eye sees only the right micrograph and the left eye sees only the left micrograph. Such stereo separation has been obtained by using red and green light or orthogonally polarized light (McKay, 1951). Some methods use a stereo projector to superimpose the images on a screen; other methods use a standard slide projector and slides which contain either superimposed or vertically separate stereoscopic images.

STEREO METHODS USING ORTHOGONALLY POLARIZED LIGHT

Polarization Method

The light that radiates from an incandescent lamp does not have a set polarization direction. When such randomly polarized light passes through a polarizing filter, it emerges with a preferred polarization direction. The filter absorbs all light that is not polarized parallel to the filter's polarization direction. Maximal absorption occurs for light polarized perpendicular to this direction. This unique property of light and of polarizing filters makes possible the discrimination necessary to visually separate the two members of a stereopair. An article by Thomas and Lentz (1972) contains a good discussion of electron stereoscopy and slide preparation and projection using the polarization method. In addition, the article lists several sources for stereoscopic viewing and projection equipment.

Polarization stereo slides can be prepared from two transparencies and projected with a stereo projector. Alternatively, a single transparency can be prepared which contains both members of a stereopair superimposed and orthogonally polarized. This slide can then be projected with an ordinary slide projector.

Both polarization methods require the use of either a lenticular screen or a flat silvered screen to maintain the polarization directions of the projected stereopair. The polarized projected images are viewed with Polaroid glasses in which the polarization directions of the filters over the two eyes are mutually perpendicular.

Glass-beaded screens lead to a loss of discrimination between the individual members of a stereopair, because the beads cause the polarized light to reflect many times before leaving the screen's surface; these reflections randomize the polarization directions of a projected stereopair so that the viewing glasses cannot separate the left image from the right.

Stereo Slides Using a 35 mm Format

The most frequently used polarization method makes use of commercially available 1⅞ x 4-inch cardboard mounts which have two square holes, one for each member of a stereopair. The two images of a stereopair are recorded on separate pieces of 35 mm transparency film; the film is cut; the left image film is taped over the left cutout in the cardboard mount, and the right film is taped over the right cutout. The two film images must be meticulously aligned vertically, horizontally, and rotationally to give the best stereo effect; alignment jigs are commercially available (Polysciences, Inc.). The horizontal alignment is less critical than either the vertical or the rotational, because the human eye can accommodate some degree of horizontal displacement of stereo images and still perceive a stereo effect.

Stereo Slides Using a 3-1/4 x 4-inch Format

Polarization stereo slides using 3¼ x 4-inch transparencies can be projected using an overhead mirror stereo projector (Buhl Optical Company). The slides are made as described above except that the film and mounts are larger, and the correct superpositioning of the stereo images is done on the projector while viewing the images projected onto a screen. The projector has alignment pins which correspond to a set of holes in the film mounts; once a slide is made, it can be placed on the projector at a later date and will normally show a good stereo image.

Stereo Slides Using the Vectograph Method

Polaroid stereo slides can be prepared by the Vectograph process; these slides do not require the use of a stereo projector. They contain a single piece of film with the two images of a stereopair separately recorded on opposite faces. These two images were processed so that the polarization direction of light transmitted through one of them is perpendicular to the polarization direction of the light transmitted through the other. When a Vectograph slide is projected onto a lenticular screen, it gives an orthogonally polarized pair of images which appear three-dimensional when viewed through the usual Polaroid glasses (McKay, 1951).

Stereo Optical Company is the only vendor which commercially handles the Vectograph process. There is no cost advantage to processing the slides oneself, since the cost of the special transfer dye and film base is nearly equal to the cost of having the slides professionally done. A 5 x 7 Vectograph print costs approximately $100 for the first slide and about $5 for every copy (Walter Lewis, Stereo Optical Company, personal communication).

The cost per slide can be greatly reduced by dividing the 5 x 7 area among several stereopairs. There are some difficulties encountered as the number of stereopairs in a given area is increased. A loss of image sharpness occurs when projecting smaller slides. The transfer dye is sensitive to the excessive heat generated by smaller format projectors not equipped with cooling fans. This dye is also sensitive to high humidity. Despite these drawbacks, polarization stereo images using the smaller slide sizes can be obtained without the use of a stereo projector (Walter Lewis, personal communication).

The procedure for making a pair of black and white transparencies which can be used in the Vectograph process is involved but straight-forward. One of the members of the original stereopair must be copied as a mirror image, because the Vectograph process uses the emulsion side of the pair of transparencies to contact-print opposite faces of the special film base. The pair of transparencies must be properly super-imposed and aligned by registration marks so that the Vectograph slide will give a good stereoscopic image when projected and viewed. Since the two black and white transparencies in themselves do not yield a stereo image when superimposed and viewed over a light box, another system which does give stereo must be used to set the alignment marks.

One suggested method involves the superposition of a red and a blue piece of diazo film, each of which contains the image of one of the black and white transparencies; the procedure for making these diazo prints is described later in this chapter. The two pieces of diazo film are placed on a light box and aligned to give the best stereo effect when the com-posite is viewed with red and blue glasses. This exact position is recorded on the two pieces of film with registration marks. The diazo prints are separated, and the corresponding black and white transparency is placed on top of each so that the registration marks can be transferred. The paired black and white transparencies are now ready for Vectograph processing. This same registration procedure was employed to align the plates used to make Figs. 5–1 through 5–4.

STEREO METHODS USING RED AND GREEN (OR BLUE) LIGHT

Stereo slides can be prepared which use differently colored instead of differently polarized light to distinguish stereo images. Figs. 5–1—5–4 are examples of red-green stereopairs. Slides can be made from two transparencies mounted in 1⅝ x 4-inch cardboard mounts; they are pro-jected with a stereo projector having a red filter over one lens and a green filter over the other. An alternative procedure uses color film which has the members of a stereopair superimposed, one in red and one in green; this method uses a standard slide projector. Red-green slides do not require a lenticular screen. The stereo image is seen through red and green filters, one for each eye.

The twin transparency red-green slides are prepared in the same way as the twin transparency polarization slides are.

There are several ways of making red-green superposition slides:

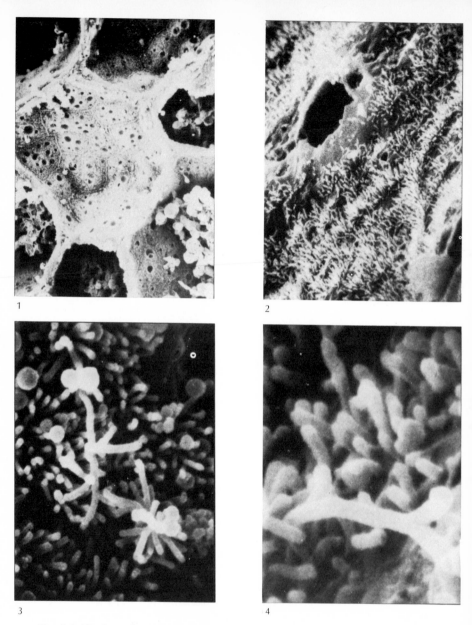

1

2

3

4

Fig. 5–1. The luminal surface of several alveoli from a 13-day lactating mouse mammary gland. Individual cells have polygonal borders. The membrane remnants of milk lipid droplets are seen in four alveoli. 550 X. All the stereomicrographs are to be viewed with a red filter over the left eye and a green or blue filter over the right eye. (See frontispiece for color.)

Fig. 5–2. The luminal surface of a mid-lactating alveolus showing three cells. Dense microvilli cover the cell surface and stand erect along cell borders. 5,200 X.

Fig. 5–3. Microvilli on the luminal surface of a lactating alveolar cell. The cluster in the upper right is 1.6 μm high. The spheres are milk protein granules. 22,000 X.

Fig. 5–4. Microvilli on the luminal surface of a lactating alveolar cell. The forked microvillus is 1.8 μm long and .1 μm wide. 52,000 X. Measurements were made as described in Nemanic and Pitelka (1971).

The Direct Method

Boyde (1971b) reported a method for making stereoscopic prints by an anaglyph* technique directly from the record cathode ray tube (CRT) of a scanning electron microscope. In his method, he suggested the replacement of the usual blue phosphor record CRT with one having a white phosphor. The members of the stereopair were superimposed on Polacolor Polaroid film (or Kodachrome IIA), with a red filter used for the first image and a green filter used for the second.

The advantage of this method with the use of Polacolor film is that since the superimposed stereopair is made directly off the microscope record CRT, errors in superposition and exposure can be identified and corrected while using the microscope. The disadvantages lie in the expense of the white phosphor CRT and the cost of extra microscope time needed to retake improperly superimposed stereo images.

Boyde's procedure can be used with the blue phosphor CRT standard on most scanning electron microscopes even though the phosphor has a poor yield of red and green light relative to the blue. The low levels of red and green light can be compensated for by increasing the brightness of the record CRT as long as special care is taken to return the scan line to the top of the CRT after each exposure. This increased brightness is sufficient to burn the tube phosphor in a short period of time if the CRT electron beam continuously irradiates the same line on the tube face.

Indirect Method Using 120 Color Film

Nemanic (1972a) reported a method which uses the standard black and white positives taken with a scanning electron microscope as a starting point for making stereo slides. The key to the procedure lies in obtaining a black and white stereopair which can be properly superimposed onto color film. The members of the stereopair must be at the same magnification and must be centered on the same spot. The z-axis is used to focus the image after tilting, and the x- and y-axis controls are used to recenter the area shown in the first member of the stereopair.

A Polaroid MP-3 Industrial Viewer, a Graflex 120 roll film back, and a framing table (or equivalent equipment) can be used to make red-green stereo lantern slides from stereopairs by the following procedure:

Place the lower angle print of the stereopair on the framing table,

---

* Anaglyph describes a stereoscopic picture in which the right component, usually red, is superimposed upon the left component in a contrasting color (usually bluish-green) to produce a three-dimensional effect when viewed through corresponding colored filters.

**Table 5–1   Exposure Using Four 150-Watt Flood Lamps
on a Polaroid MP-3 Industrial Viewer**

	Shutter Speed	*Lens Opening*	
Green (Wratten 58)*	1/2	4.7	Ektacolor Type S (ASA 100) (for negative slides)
Red (Wratten 25a)	1/8	5.6	
Green (Wratten 58)	1/2	5.6	High Speed Ektachrome (ASA 160) (for positive slides)
Red (Wratten 25a)	1/8	8	

* It is convenient to mount the filters between 2 x 2-inch slide glasses bound together with aluminum tape.

center the micrograph by viewing it through the camera lens, and enlarge the image to fill the roll film back's format. Place Ektacolor color film in the roll film back (Ektachrome may be used to make positive slides, but negative slides give a better stereo effect). Expose the first member of the stereopair with the red filter over the lens (see Table 5–1); for consistency, standardize the use of the red filter for the lower angle micrograph. Without advancing the film, place the second micrograph in the same frame as the first; since both original micrographs of the stereopair are centered on the same specimen area, their picture border, not their photographic information, is used to align the micrographs and assure the correct superposition of images on the color slides. Set the shutter speed and lens opening for the green filter (see Table 5–1); expose and advance the film for the next slide.

Have the exposed 120 film commercially developed and returned unmounted. Place each color negative between 3¼ x 4-inch lantern slide glasses with the tilt direction of the stereopair parallel with the long axis of the lantern slide glass. Mask off the clear glass with black tape and bind the slide glasses together with aluminum tape.

It should be noted that the black and white micrographs of each stereopair must be carefully inspected when they are taken, to ensure that they are exactly centered on the same spot. Sometimes there is a small amount of image shift after the second image is centered and before the micrograph is taken; this shift is especially troublesome at higher magnifications (e.g., greater than $\times 20,000$).

The advantages of this method are: it does not require a white phosphor cathode ray tube; it uses standard photocopying equipment; and it is applicable to both scanning and transmission electron microscopy.

Indirect Method Using Diazo Film

McGee-Russell and Speck (1970) have used diazo material (Technifax Corporation) to make color transparency prints from electron micrograph negatives. The negatives were taken as a tilt series rather than as a stereopair; the series consists of negatives corresponding to $+6°$, $0°$, and $-6°$, or preferably as many as five negatives exposed at $+6°$, $+3°$, $0°$, $-3°$, and $-6°$. Each negative was copied on Kodalith high-contrast positive transparency film; the positive film was placed in a pressframe with a sheet of red, green, or blue diazo film; red was used for $+6°$, $+3°$, and $0°$, and green or blue was used for $0°$, $-3°$, and $-6°$. Each diazo-Kodalith combination was exposed for several minutes to a UV lamp and developed in warm ammonia vapors.

The diazo prints from each positive Kodalith print of the tilt series were sandwiched together and superimposed to give the best stereo effect. The composite was illuminated from behind and copied onto 35 mm color slide film. A forthcoming paper will give a more detailed description of the technique (S. M. McGee-Russell, personal communication).

Superposition matching of widely separated stereopairs (12 to 14°) usually gives a somewhat distorted stereoscopic result, depending on the depth seen in the image. The intermediate tilt angles tend to compensate for this and improve the information content of the final image.

Unlike the two previous red-green stereo methods, this procedure does not require that the members of the stereopair be centered on the same area, because the superposition of images is done using separate pieces of film over a light box. This method is also very inexpensive because diazo film costs only a few cents for an 8½ x 11 sheet.

Modified Diazo Method

In the author's opinion, a modification of the method of McGee-Russell and Speck has proved to be the best general procedure for making stereo slides from electron micrograph stereopairs.

Since the exposure of the diazo film by ultraviolet light involves a breakdown of one of two dye precursors in proportion to the amount of light transmitted through the overlying positive transparency, diazo contact transparency prints display good continuous tone properties. Stereopairs in the scanning microscope are routinely taken with a 7° tilt difference; this angle is sufficient to give good three-dimensional images of surface detail.

A variety of positive or negative black and white films can be used to take the stereopairs; two such possibilities are Polaroid type 46L and Polaroid P/N-55; the exposed film should have its width parallel with the stereo tilt axis, and it should not show extremes of contrast. The positive or negative film is placed over red or green (or blue) diazo film, emulsion side to emulsion side, and the composite is exposed to ultraviolet light; two 15 watt fluorescent UV lamps at a distance of 18 in. give a complete exposure in 5 to 10 min. The exposed diazo film is then developed in warm ammonia vapors ($\sim40°C$); red diazo takes 30 to 60 sec; green, blue, and cyan take much longer (5 to 10 min). The red and green diazo films are superimposed to give a good stereoscopic effect when viewed over a light box through red and green glasses. The composite is placed between 3¼ x 4-inch lantern slide glasses and bound with metal tape.

Since the diazo stereopair is the same size as the Polaroid film from which it was made, it nearly fills the 3¼ x 4-inch slide glass with photographic information; almost half of a slide made with 120 film is unused. Furthermore, the original black and white prints need not be centered on the same area, although the resultant film overlap will reduce the size of the stereo image in the slide. Since the diazo film is sensitive only to ultraviolet light, a photographic darkroom is not necessary for slide processing; in addition, the equipment needed costs very little (about $20).

Viewing Glasses

There are several sources for red and green filters suitable for viewing stereo slides. Red-green glasses may be purchased from Bernell Corporation. Red and green acetate found in an art supply store can be used to make inexpensive, hand-held filters for large groups of people; two thicknesses of green and three of red approximate the color of optical density of the Wratten filters used to make color slide composites. A third, very inexpensive, alternative involves the use of unexposed red and blue diazo film; the film is developed in warm (40°C) ammonia vapors to give suitable sheets of red and blue acetate.

THE NESCH VERTICAL SYSTEM

Mannheim (1971) described a method of stereo slide presentation which projected both members of a stereopair onto an ordinary screen with one image above the other, not overlapping. The stereo effect came from viewing the screen with plastic glasses which had a separate prism for each eye to bring the images together. The Nesch method also provides an alternative method for presenting stereo images in publications, as long as the prism glasses are available (Stereo Vertrieb Nesch).

The advantages in the method are that it does not require special slide preparation or a stereo slide projector; it also can present stereo images in their natural colors. The disadvantage is that there is a set radius and solid angle from the center of the screen for the best stereo effect. This severely restricts the seating capacity and arrangement of an audience attending a stereo presentation, but it does not interfere with viewing stereopairs printed in journals.

COMPARISON OF THE METHODS

Polarization Method

Polaroid stereopairs can present black and white or natural color stereoscopic images. These slides also can display more levels of contrast than can be seen in red-green stereo slides, because the latter must be printed with intermediate contrast so that dark green detail does not mask the light red. In addition, polarization stereo slides offer better image separation than red-green slides (Boyde, 1971b). The slides are easy to make, and the alignment jigs, which are commercially available, can minimize the amount of projector realignment needed when showing a series of slides.

There are many disadvantages in the polarization method. It requires a stereo projector (this does not apply to Vectograph slides), which is expensive and not commonly available in lecture rooms. The method also requires a lenticular screen to maintain the slides' orthogonal polarization directions. This screen must be stretched flat; otherwise light and dark spots will appear in the stereo image. Because the demand for lenticular screens is low, they are usually small by auditorium standards. Furthermore, the polarizing filters in the stereo projector reduce the brightness of each image nearly four times; the use of larger screens would require more powerful projection equipment.

Some skill is required to project and properly superimpose the number of polarization stereo slides used in a stereo presentation, because in addition to the focus controls, a stereo projector has a separate pair of adjustments for vertical and horizontal image alignment. Lacking an experienced projectionist, a speaker frequently finds it necessary to project his own slides. Finally, Polaroid viewing glasses cost fifty and sixty cents each.

Red-Green Method

This method uses a standard projector and does not require a lenticular screen. Since the superposition red-green slides are prealigned, there is no need for readjustment of the projected stereo image from slide to slide. Red and green (or blue) filters can be made from inexpensive materials. If color film is used to make the slides, prints can be made at the time of commercial film development; these prints provide a convenient means for presenting stereoscopic images to small groups of people, who simply view the composite through red and green filters.

The disadvantages of the red-green method are that the best stereo effect is achieved with a negative rather than a positive image; the reverse contrast combined with the structural complexity inherent in some slides sometimes requires an extra amount of time for viewers to adjust to what they see. In these cases, it is advisable to precede a color stereo slide with a black and white slide of the same area. The other drawbacks of red-green stereo slides are indicated in the section on polarization slide advantages.

CONCLUSIONS

Most of the difficulties in using the polarization stereo method will vanish when the demand for stereo presentation increases and the equipment becomes more commonly available. The polarization method is generally held to be superior to the red-green method. In the meantime, the red-green superposition methods provide a useful alternative, since they require no image realignment, special screen, or stereo projector.

I would like to thank Thomas Hayes, James Pawley, S. M. McGee-Russell, Alan Boyde, and Dorothy R. Pitelka for their helpful suggestions in writing this paper. I would also like to thank Thomas Everhart for allowing me the use of his Cambridge Stereoscan Mark IIa Scanning Electron Microscope (purchased under GB-6428 N.S.F. and maintained by GM-17523 N.I.H.). In addition, I would like to express my appreciation to John Underhill for his help with photography and to Reiko Kubota and Janie Beverly for typing the manuscript. The author is supported by USPHS grants CA-05045 and CA-05388.

APPENDIX

Commercial Vendors

Bernell Corporation, 316 South Eddy Street, South Bend, Indiana, 46617.
Buhl Optical Company, 1009 Beech Avenue, Pittsburgh, Pennsylvania, 15233.
Polysciences, Inc., Paul Valley Industrial Park, Warrington, Pennsylvania, 18976.
Stereo Optical Company, 3589 North Kenton Avenue, Chicago, Illinois.
Stereo Vertrieb Nesch, 44 Munster, Enschedewey, 78, West Germany.
Technifax Corporation, Holyoke, Massachusetts.

REFERENCES

Boyde, A. (1970). Practical problems and methods in the three-dimensional analysis of scanning electron microscope images. In: *Scanning Electron Microscopy. Proc. 3rd Ann. Scan. Electron Micros. Symp.*, p. 105. I.I.T. Research Institute, Chicago.

Boyde, A. (1971a). A review of problems of interpretation of the scanning electron microscope image with special regard to methods of specimen preparation. In: *Scanning Electron Microscopy. Proc. 4th Ann. Scan. Electron Micros. Symp.*, p. 3. I.I.T. Research Institute, Chicago.

Boyde, A. (1971b). Recording anaglyph stereopairs in the SEM and some other uses of color images in the SEM. *Beitr. Electronenmikroskop. Direktabb. Oberfl.* **4/2**, 443.

Hama, K., and Nagata, F. (1970). Stereoscopic observations of biological specimens by means of the high voltage electron microscope. In: *Microscopie Electronique*, Vol. 1 (Favard, P., ed.), p. 461.[*]

Howell, P. G. T., and Boyde, A. (1972). Comparison of various methods for reducing measurements from stereo-pair scanning electron micrographs to "Real 3-D Data." In: *Scanning Electron Microscopy. Proc. 5th Ann. Scan. Electron Micros. Symp.*, p. 234. I.I.T. Research Institute, Chicago.

McGee-Russell, S. M., and Speck, R. (1970). Color transforms and color fusion stereoscopy for the study of transmission tilt, high voltage and scanning electron micrographs, and freeze-etch replicas. *J. Cell Biol.* **47** (2 part 2), 133a.

McKay, H. C. (1951). *Three-Dimensiosal Photography: Principles of Stereoscopy.* Jones Press Inc., Minneapolis.

Mannheim, L. A. (1971). A steroscopic system for any size picture pairs (Nesch Vertical System). *Photographic Application in Science, Technology and Medicine* **38**, November issue.

Nemanic, M. (1972a). Preparation of red-green stereo lantern slides from SEM micrographs. *Proc. 30th Ann. Meet. Electron Micros. Soc. Amer.* (Arceneaux, C., ed.), p. 412.

[*]Société Française de Microscopie Electronique, Paris.

Nemanic, M. (1972b). Critical point drying, cryofracture and serial sections. In: *Scanning Electron Microscopy. Proc. 5th Ann. Scan. Electron Micros. Symp.*, p. 297. I.I.T. Research Institute, Chicago.

Nemanic, M., and Pitelka, R. (1971). A scanning electron microscope study of the lactating mammary gland. *J. Cell Biol.* **48**, 410.

Staehelin, L. A. (1970). Stereo electron microscopy applied to high resolution freeze-etch replicas. *Proc. 28th Ann. Meet. Electron Micros. Soc. Amer.* (Arceneaux, C., ed.), p. 306.

Thomas, L. E., and Lentz, S. (1972). Preparation of stereo slides from electron micrographs. *EMSA Bulletin* **2**(**2**), 10.

6. LOW-MAGNIFICATION STUDY OF UNCOATED SPECIMENS

H. F. Howden and L. E. C. Ling

Biology Department, Carleton University, Ottawa, Canada

INTRODUCTION

The major impact of scanning electron microscopy has been in the field of high magnification, with recent developments tending to emphasize higher vacuums and higher resolutions. Only desultory investigations have appeared dealing with low (1 to 600×) magnifications (Howden and Ling, 1973), although scanning electron microscopes have an obvious advantage, because of their large depth of focus, over optical systems.

Recently there has been an increase in the application of scanning electron microscopy to biology, but most studies have been limited for several reasons. Since most biological materials are largely nonconductive, samples have usually been coated with gold or other conductive materials. Reasons for this are well documented, the coating being necessary to bleed off excess incident electrons that rapidly accumulate as a result of voltages (5 to 55 kV) needed to obtain satisfactory signal levels (i.e., signal-to-noise ratio) at high magnifications.

However, there are many types of specimens and situations where such magnifications are not needed and where it is highly desirable not to alter the specimen by using a metallic coating. Examples of this are numerous. Many specimens, both plant and animal, can best be described through illustrations, but with few specimens available it is often not desirable to coat them. In legal cases, where it is often of interest to study bones, hair, paper, and so on, coating specimens may be considered undesirable because the evidence is changed and is therefore inadmissible. This chapter deals primarily with the problems of working at low magnifications using uncoated specimens. Also discussed are methods of handling material and the types of microscope modification that may be needed in order to obtain reasonable results. The latter aspect will be dealt with first.

MODIFICATIONS OF THE INSTRUMENT

In the 1972–73 *Science* "Guide to Scientific Instruments" issue, there is a list of thirteen SEM manufacturers who have a variety of models for sale in the United States, however this does not give any indication of what is available on the world market. We have not been able to examine all the machines, and with many changes being made every year, the types of modification needed for any particular model will vary. For all the models we have seen to date, some modifications will probably be desirable in order to obtain the best results when using uncoated, nonconductive samples.

In general, a variable working distance up to 50 mm is needed with a power source that can be accurately controlled and calibrated between 1 and 5 kV. No machine that we have seen with a single power source extending from 1 to over 30 kV meets the latter requirement; however, this may change. We are not familiar with many models, and the discussion of modification and specimen handling is based on the work carried out with a JEOL JSM-U3.

Essential modifications encompass three categories: power sources, an additional supplementary scanning coil, and changes to the specimen chamber or holder. On our machine, an additional high-voltage (0 to 5 kV) power supply was installed (a Fluke model 408B), which allows voltage changes in .05 volt increments or more. This has proved very satisfactory, but a larger model giving a supply from 0 to 10 kV would be occasionally useful, although we rarely have had a need to exceed 5 kV. At present, for power in excess of 5 kV, one must change the main power cable of the gun over to the U3 power supply, which ranges from 5 to 55 kV. The changeover requires approximately five minutes. Also, for better contrast, the electronic circuitry affecting the "Gamma Control" was modified (Kimoto, 1972, p. 24).

While working with low voltages of 1.5 to 3 kV, it was difficult to maintain a sufficiently tight incident electron beam (focus) when using working distances up to 50 mm. To overcome this, the focusing (objective lens) circuitry was modified. This enabled a lowering of the focal point of the incident beam at a lower kV. The technique of double focusing, using the coarse objective lens setting of ~10, was not found satisfactory.

In order to obtain low-magnification micrographs of large objects (2 cm) at either 50 or 32 mm working distance, a specially wound supplementary scanning coil was installed. This coil was placed at the bottom of the objective lens (top of the specimen chamber), and can be electrically cut in or out of use without affecting normal operation.

The supplementary scanning coil, situated at the bottom of the objective lens, further enlarges the scanning area, since the incident beam is not limited to the diameter of the internal objective lens configuration. When a higher magnification requires working with coated specimens, the 13 mm working distance can be used by removing the supplementary scanning coil, which requires ~10 min to complete. This enables the

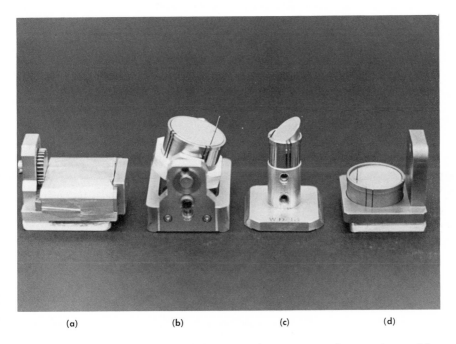

(a)   (b)   (c)   (d)

Fig. 6–1. Substage and specimen holders: (a) 4 cm^2 low (10 mm) substage with central indentation filled with low-vacuum embedding clay covered with aluminum foil; the substage has a 15 mm lateral movement to either side. (b) Large, variable height holder with soldered pin and 90° tilt substage. (c) Small, 45° fixed tilt, variable height holder with fixed substage. The 45° tilt, when coupled with the goniometer tilt (+45°, −5°) and rotation, allows variable specimen tilting from 0° to 90°. (d) Large, variable height holder with soldered pin and fixed low (10 mm) substage.

specimens to be situated in the high-magnification position, thus improving resolution.

In many machines, low magnifications (below 20×) usually necessitate some changes in either the specimen holders or the substage. In the U3, the goniometer, when set at the lower working level, allowed a 50 mm working distance if holders did not exceed 10 mm in height. Several types of holders were found useful, most being easily manufactured in any well-equipped science workshop. Fig. 6–1 illustrates some of the types that can be used in the U3. The major limitations are the size of the holder that can be introduced into the chamber and the type of holder needed to best handle the object being studied. Adequate holders and ease of positioning the specimens are extremely important, since uncoated specimens should be exposed to the scanning beam for as brief a duration as possible, in order to prevent charging.

PREPARATORY PROCEDURES

We have used dead, dried insects (especially beetles) that are either pinned with a steel insect pin or, in the case of small specimens, glued to a small paper point that is attached onto the right side of the insect, a pin being run through the other end of the point. The specimen is first placed in a humidor for 24 hr (this step can be omitted if material is plentiful) and then cleaned in a fairly powerful ultrasonic cleaner (e.g., Electromations Ultrasonic Cleaner, model LP-1HD).

Specimens are pinned directly into a holder filled with low-vacuum embedding clay with the surface covered with aluminum foil. Specimens on points are left with the pin through the point, or the point and specimen are transferred to a very short pin soldered to the holder. No attempt is made to render the paper points conductive, and no effort is made with either pinned or pointed specimens to "ground" the specimen. If the specimen is not pinned or pointed, it can be attached to the holder with glue, double-sided Scotch tape, silver paint, etc. A very small quantity of glue or paint should be used, as it may spread over the specimen or cause unwanted charging.

After being attached, the specimen in the holder is positioned for the desired view under an optical dissecting microscope. This is done in order to save positioning time in the SEM chamber, thus reducing the duration of exposure to the beam. Once one is familiar with the type of material being handled, the time saved in outside positioning is an important factor in reducing the charging. The specimen is then placed in the goniometer of the SEM. In the JEOL JSM-U3, specimen size is normally limited for practical purposes by the size of the "air lock" (48 x 62 mm). The goniometer rotates 180° in each direction, tilts +45° to −5°, with lateral (X) movements of ±5 mm and forward (Y) movements of ±12.5 mm. There are two fixed working distances (32 and 13 mm), but we vary this by utilizing a variety of holders of different heights or by placing the insect pins at different heights.

With the specimen in the chamber and proper vacuum attained, the beam is turned on and the area to be viewed positioned as rapidly as possible. We use a TV attachment for this, considering it slightly better than a rapid scan. After positioning, we switch to a 10 sec scan for focusing. Normally 1.5 kV is used, the current being increased in 100 volt increments if necessary. If the area being studied is too dark or too bright in relation to other areas, or if there is a moderate amount of charging, some rotation of the specimen will often give a radical improvement.

With difficult specimens, blanking the beam for a few minutes occasionally is helpful. Our pictures are normally taken at a scanning speed of 50 sec. Altering the scanning speed may be useful—that is, using a 25 sec scan, which decreases the time that the incident beam remains on a single spot. This technique reduces charging, but detected information is also decreased. We found this most useful in habitus shots of some beetles, using different types of film for greater or lesser contrast; however, we have not fully investigated these aspects as yet. In summary, typical settings for low kV micrographs using our JSM-U3 are: voltage, 1.7 kV, PMT, 3; collector, full voltage; condensor, 1 or 2; objective, 1; and scanning speed, 50 sec. These settings will vary depending upon the instrument used; for example, the signal-to-noise ratio will determine the PMT and the collector settings.

The methods outlined above are generally successful, as can be seen in Figs. 6–2 and 6–3. However, certain types of specimens are almost impossible to study without coating. Some insects and moths, for example, are covered with scales which are usually shed during pumpdown and which contaminate the barrel. It is possible to obtain micrographs of uncoated specimens of this type, but this necessitates cleaning the machine several times a day.

Smooth, shining, very convex beetles are extremely difficult to photograph at low magnifications (below 50×) even when coated. Some freshly killed soft-bodied insects may expand under vacuum or, conversely, if allowed to dry, may collapse. To avoid this, freeze-drying may be needed. However, a vacuum evaporator set at moderate vacuum can dry fairly large soft-bodied forms. The specimens are kept under vacuum

(a)   (b)

Fig. 6–2. Uncoated, pinned beetles: (a) Head and pronotum of *Onthophagus concinnus* Lap., 2.2 kV. 22 X. (b) Head of *Bolbothyreus ruficollis* (Br.), 2.2 kV. 45 X.

(a)   (b)

Fig. 6–3. Fresh, uncoated geranium flower: (a) Central 5 mm of a flower, 2.1 kV. 20 X. (b) Portion of an anther showing pollen grains, 2.5 kV. 250 X.

in the evaporator for 24 to 48 hr. The vacuum is then slowly released, and the majority of specimens retain their shape and can be handled in the SEM as described above.

PROBLEMS

Handling of biological materials varies somewhat with the type of specimen (Russ and Kabaya, 1971; JEOL Application Department, 1972), but working with any nonconductive material involves problems with charging. Before specific examples are given, some guidelines can be mentioned. As low a kV as possible commensurate with satisfactory micrographs should be used, since the lower the beam current, the slower will be the charge buildup (Kosuge *et al.*, 1970). Exposure to the scanning beam should be limited to as brief a duration as possible. Sometimes blanking the beam by closing the gate valve after focusing for a few minutes, then opening the gate valve immediately before taking a picture, is helpful. Increasing the condensor setting just before taking the picture is also advantageous.

If possible, fresh specimens (both plant and animal) that contain some moisture should be used. The gradual release of water molecules (or a film of water) probably reduces charging by bleeding off some of the electrons. Moisture may be replaced in a dry specimen by placing it in a humidor for 24 hr. Even slightly wetting the surface of a specimen before placing it in the chamber is helpful. Brody and Wharton (1971) found a mixture of glycerol (96.6%), potassium chloride (0.05%), and water (3.35%) useful in reducing the charging when studying certain species of mites.

However, care should be taken with these techniques, since large amounts of water or other volatile substances may, in the long run, be damaging to the pumping system of the machine. Sometimes simply changing the position of the specimen is sufficient. We have had reasonable results utilizing at least some of the above procedures when studying teeth, bones, and some plant material, but most of our work has dealt with insects.

The use of the SEM to obtain low-magnification micrographs presents at least one other problem in addition to charging. The problem is one of distortion, which, in the work described above, is of at least two types. The first type involves inherent machine distortion evident when studying a flat surface, such as a test screen (Fig. 6–4). In our U3, at a magnification of approximately 10×, the measurable distortion is ~3%. This type of distortion is probably due to the method of winding the lens coils.

A second type of distortion occurs when studying convex or concave

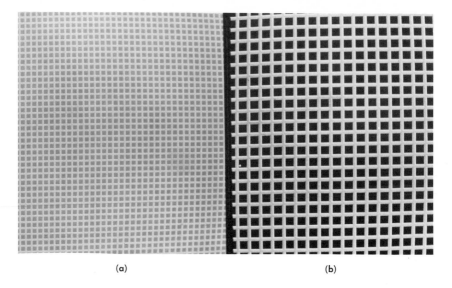

(a)   (b)

Fig. 6–4. Flat metallic test grid, 8 squares equal 1 mm: (a) Micrograph taken at 1.9 kV; distortion most pronounced toward right side. 20 X. (b) Micrograph taken at 1.9 kV; distortion as in (a). 40 X.

objects at low magnifications, and is caused by large differences in the height of the area being scanned. Since the SEM magnification is directly related to the distance of the subject from the scanning coil, tilt or unevenness will produce some distortion; for instance, tilting a flat test screen will give a trapezoidal effect, and tilting a convex or concave subject will often add to the distortion. A convex-concave type of distortion is illustrated in Fig. 6–5. The two micrographs were taken in the same position (no tilt) and at the same magnification (height).

At present, a ready solution to the distortion problem is not available. The problem can, however, be lessened in two ways. If the specimen fills the picture frame at 20×, lower the magnification to 12× or 5×. This can often be accomplished by simply lowering the specimen. This procedure slightly lessens the variation in the angle of the beam, since the beam is not scanning the subject near the periphery. The second method is to increase the magnification significantly so that only a portion of the specimen is being scanned. A mosaic of four or more micrographs can be combined into an enlarged micrograph of the specimen. This will succeed if the specimen is not tilted or rotated. Thus distortion is usually decreased, since by increasing the magnification significantly and scanning a small part of the specimen, the curvature covered by the scan for a single picture is usually lessened.

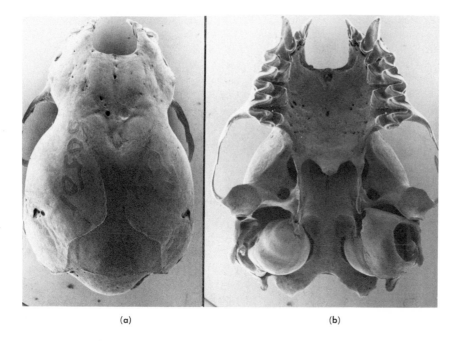

(a)   (b)

Fig. 6–5. Untreated skull of a bat, *Glauconycteris argentata* (Dobson); zygomatic width 9.3 mm., length from inner incisor to occipital condyle 12.3 mm, 1.9 kV. 10 X. (a) Dorsal view with convex type of distortion. (b) Ventral view with concave type of distortion. The pictorial difference in the width of the zygomatic measurement is due to both the convex and the concave effects and to the different distances from the lens of the zygomatic arches (affecting magnification) when the skull was turned over without changing the height of the holder. If the holder had been lowered to pictorially equalize the zygomatic breadth, the length would be equally out of proportion.

Generally, at low magnifications some distortion occurs. In the area of usage of the SEM for low-magnification studies, there is an obvious need for further research in lessening the charging and distortion. Also, some aspects of the above problems may be eliminated by improving the photographic equipment and techniques.

This work was supported by grants from the National Research Council of Canada.

REFERENCES

Brody, A. R., and Wharton, G. W. (1971). The use of Glycerol-KCL in scanning microscopy of Acari. *Ann. Ent. Soc. America* **64**, 528.
Howden, H. F., and Ling, L. E. C. (1973). Scanning electron microscopy: Low-magnification pictures of uncoated zoological specimens. *Science* **179**, 386.

JEOL Application Department (1972). Preparation of biological specimens for scanning electron microscopy. *JEOL News* **10e**, 42.

Kimoto, S. (1972). The scanning microscope as a system. *JEOL News* **10e**, 2.

Kosuge, T., Hashimoto, H., Sato, M., and Kimoto, S. (1970). Quality of the secondary electron image at low accelerating voltage. *Proc. 28th Ann. Meet. Electron Micros. Soc. Amer.* (Arcenaux, C. J., ed.). Claitor's Publishing Division, Baton Rouge, La.

Russ, J. C., and Kabaya, A. (1971). Preparation of samples for scanning electron microscopy. *JEOL News* **9e**, 27.

7. SPORES

Ann W. Nickerson, Lee A. Bulla, Jr., and Cletus P. Kurtzman

Northern Regional Research Laboratory, Agricultural Research Service,
U.S. Department of Agriculture, Peoria, Illinois

INTRODUCTION

Although the light microscope is satisfactory for examining some surface features of spores, an accurate description of surface features requires a resolving power beyond its capability. Certainly the light microscope cannot provide the magnification needed for detailed examination of spores of the actinomycetes and other bacteria. Although the transmission electron microscope (TEM) has adequate magnifying and resolving powers, its usefulness is limited in the study of overall morphology. The preparation of satisfactory carbon replicas and freeze-etchings of highly ornamented spore surfaces can be technically difficult, if not impossible.

The scanning electron microscope (SEM), with its wide magnification range (20× to 50,000×), is ideally suited for the study of spores. Bulla *et al.* (1973) have indicated that the SEM is a powerful tool for a variety of studies of microorganisms. Used in conjunction with the light microscope and TEM, SEM can provide concise information on spore morphology.

Relatively large areas of the specimen (∼1 cm²) can be examined with the SEM, which allows reliable estimation of morphological variations in the given specimens. Accommodation of specimens of large size renders the instrument suitable for developmental studies of the sporulation process in fungi and actinomycetes (Hawker, 1968; Williams and Sharples, 1970; Cole and Aldrich, 1971). The same feature of the instrument can be employed for studying the infection and invasion of natural hosts by fungal pathogens (Locci, 1969a and c; Locci and Bisiach,

1970 and 1971; Lewis and Day, 1972). Surface ornamentation of fungal spores, both vegetative (e.g., conidia) and sexual (oospores, zygospores, ascospores, and basidiospores), is easy to examine with a SEM; spore surface features can serve as major criteria for delimiting species, genera, and even families of fungi.

In addition to its value in taxonomic and developmental studies, the SEM can be an important tool for physiological research. Surface changes that occur during spore germination and subsequent outgrowth can be followed in yeasts (Rousseau et al., 1972) and bacteria (St. Julian et al., 1971; Ellis et al., 1971). The changes in bacterial cell surfaces following treatment with antibiotics and other surface-active chemical agents can be demonstrated (Greenwood and O'Grady, 1969; Klainer and Perkins, 1971; Watanakunakorn et al., 1971).

The SEM is amenable to a wide range of studies on microorganisms in their natural habitats. Phenomena of possible significance to the survival of microorganisms in nature, such as perforation of conidial walls during dormancy in natural soil, can be observed (Old and Robertson, 1969). The design of the instrument permits microbial colonies to be examined in situ on diverse substrata ranging from soil (Locci, 1969b) and plant surfaces (Barnes and Neve, 1968; Leben, 1969) to miscellaneous deteriorated materials (Locci, 1971); detailed observations of deterioration of both natural and manufactured materials are possible. Such investigations can lead to an increased appreciation of the ecological importance of the "microniche."

MINIMAL SPECIMEN PREPARATION

Artificial changes in microbial specimens during preparation for microscopy are not uncommon. To minimize such artifacts, the microscopist should have a general knowledge of the characteristic growth pattern and morphogenesis of the microorganism under investigation. Whenever spores of actively growing cultures are to be examined, factors such as temperature of incubation, nutrients, aeration, and illumination must be taken into consideration; all of these parameters can significantly affect cellular growth, form, and structure. Awareness of all these effects is indispensable for proper evaluation of the final electron micrograph.

Different types of spores exhibit considerable variation in their physical properties, especially rigidity. Even spores of closely related organisms may differ considerably in their amenability to similar preparatory procedures. Consequently techniques that are satisfactory for one spore type may be wholly unsatisfactory for another. Generally the nature of the organism and its dormant state as well as the objective of the study dictate the method of preparation.

Preparation of fungal spores can be simple. Often the spores are merely attached to a specimen stub, air-dried, and then metal-coated. Such minimal preparation is used with remarkable success on a diverse assemblage of organisms: fungal pathogens and saprophytes on bark fragments and other plant parts, conidia of sporogenous cultures grown on artificial solid media, and basidiospores collected from beneath mushroom caps. The spores may be dusted onto specimen stubs containing an adhesive or onto a circular glass coverslip which, in turn, is mounted on the stub; or they may be mounted by carefully placing an inverted stub on the specimen material. Spores forcibly discharged from fruiting bodies may be readily collected by appropriate placement of the stub near or under the fruiting body.

The actinomycetes may require somewhat different handling, because the form of their aerial growth is an important taxonomic characteristic for classification. A particularly useful method for observing the fragile sporulating structures with a minimum of disturbance involves growing them on partially submerged glass coverslips (Williams and Davies, 1967). The coverslips are sterilized by autoclaving and then inserted at an angle of ~45° into solid nutrient medium contained in a petri dish; approximately half of the coverslip remains exposed above the medium. A fine wire needle is used to spread an inoculum along the line where the upper surface of the coverslip meets the agar.

During incubation, the organisms grow on the medium and in a line across the upper surface of the coverslip. After mature sporulating structures have been produced, the coverslips are carefully withdrawn from the medium so that the line of growth remains attached to the coverslip. The coverslip can then be glued to a stub and examined in the SEM.

Several other methods can be used to observe microorganisms in a natural configuration. An inverted stub, with or without adhesive, may be carefully pressed onto the surface of microbial growth. The cells which attach to the stub during this procedure should reflect the natural physical arrangement of cells on a solid medium. Likewise, agar blocks containing colonies can be excised from nutrient plates with a dissecting knife, and the blocks transferred directly onto specimen stubs (Roth, 1971). Cells may also be grown on a layer of agar thinly spread over a coverslip, provided it is kept in a moist chamber. At the desired stage of growth, the coverslip is cemented to a stub and the specimen processed for microscopic examination.

The major difficulty encountered with the latter two techniques is severe shrinkage and distortion of the agar surface during evaporative metal-coating and observation under a high vacuum. This distortion may be sufficient to cause serious rearrangement of the cells from their natural state. In studies of the colonial growth of bacteria, the problems associated with shrinkage of agar surfaces can be circumvented by culturing the bacteria on an agar plate overlaid with sterilized dialysis membrane (Afrikian, St. Julian, and Bulla, unpublished data). After sufficient growth, the membrane is carefully stripped from the agar, trimmed, and transferred to the specimen stub.

In the aforementioned procedures, spores are taken directly from a growth surface and examined without cleaning. This kind of preparation has the obvious merits of ease and rapidity. However, unless undisturbed growth is of particular interest, it is advisable to include washing as a preliminary step for processing spores.

SPECIMEN CLEANING

The purpose of cleaning is to clear the spores of surface contaminants and to remove cellular debris. Surface contamination by salts, peptides, or polysaccharides is particularly serious with spores of yeasts and bacteria that have been produced in liquid culture. Such contamination may also present serious problems of interpretation for fungal spores that have been in contact with a moist substrate.

Generally spores are cleaned by washing in distilled water or a simple buffer solution. Isotonic buffer solutions are needed for specimens sensitive to abrupt changes in osmotic pressure. Both Tris and phosphate buffers are satisfactory, and can be made isotonic by addition of a low molecular weight sugar (usually sucrose). When buffered solutions are used to wash spores, final rinsing in distilled water is necessary to remove salts which may crystallize on and around the spores during dehydration, interfering with the quality of the final micrograph.

While preparing fungal conidia, it is often advisable to add a small amount of a detergent such as Teepol to the initial washing solution to facilitate wetting the conidia. Individual conidia may be separated from chains by vigorous shaking, with or without a small amount of sand added to the suspension. The sand will rapidly settle after shaking, and the spores can be easily decanted. Single conidia then can be separated from any remaining intact chains by filtering the suspension through several thicknesses of cheesecloth or nylon mesh.

Yeast and bacterial spores produced in liquid culture are normally harvested by centrifugation and washed several times with cold, sterile distilled water. For species in which the asci or sporangia lyse naturally, this procedure will produce clean spore preparations if the top layer of the pellet is discarded after each centrifugation.

Free spores of various *Bacillus* species can also be purified by a liquid two-phase partition system (Sacks and Alderton, 1961). The two-phase system involves shaking a suspension of spores with a mixture of polyethylene glycol (Carbowax 4000, Union Carbide Co.) and potassium phosphate buffer (3 M, pH 7.1). The spores will enter the upper phase, and can be recovered by centrifugation free from vegetative debris. Centrifugation should be followed by thorough rinsing in distilled water to insure that all contaminating substances are removed from the surface of the spores.

Relatively drastic preparatory methods are required for those bacteria, yeasts, and fungi that do not discharge free spores. Mere crushing of the fruiting bodies of ascosporogenous fungi by applying slight pressure to a coverslip overlaying a slide mount may be sufficient to rupture the ascus walls and liberate free ascospores. With some fungi, the walls of the asci will autolyse if the culture is allowed to age for a sufficient length of time, and the ascospores may then be harvested by flooding the growth plates with water. The resulting suspension should be shaken vigorously to facilitate release of the spores from the fruiting bodies and mycelial debris. Filtration through nylon mesh removes large debris, and differential centrifugation can be used to remove the remaining fragments.

Enzymatic digestion of the ascus may be required to obtain free spores of other ascogenous fungi, particularly some yeasts. The commercially available enzyme preparation, Glusulase (Endo Laboratories, Inc., Garden City, N.Y.) is very effective (Wickerham and Kurtzman, 1971; Kurtzman *et al.*, 1972). A small loopful of cells is mixed with 0.1 to 0.2 ml of Glusulase and incubated at 25°C. When approximately half of the spores are free (determined by light microscopy), 5 ml of cold distilled water is added and the cells are removed by centrifugation. The duration of incubation in the enzyme is species-dependent, and varies from a few minutes with *Saccharomycopsis crataegensis* to 3 hr or longer for some species of *Schwanniomyces*. Three additional washes in distilled water are generally sufficient to produce clean spores.

Bacteria such as *Bacillus thuringiensis*, with persistent sporangia, may require sonication to break the sporangium and liberate free spores (Bulla *et al.*, 1969). For this method, the spores contained within intact sporangia are harvested from the growth medium by centrifugation, washed five times in distilled water, and resuspended in thick-walled test tubes. The suspensions are then sonically vibrated while the tubes are partially immersed in an ice bath until more than 95% of the spores have been freed from the sporangia. The specific period of time required to remove the sporangia varies between 5 and 60 min depending upon the particular *Bacillus* species being treated.

If the spore suspensions are allowed to stand for 20 min after sonication, much of the sporangial debris will settle to the bottom of the tube. The free spores that remain in suspension can be withdrawn using a Pasteur pipette. After three additional washes in distilled water, the spore suspension is shaken in a 1% solution of sodium dodecyl sulfate for 10 to 15 min at 37°C to reduce clumping. The spores are removed by centrifugation and washed four times in 1 M NaCl followed by five washes in distilled water.

It must be emphasized that all steps of specimen preparation should be monitored with the light microscope to give the investigator some indication of the cleanliness of the sample and to provide early indication of any adverse effects on the specimen. Monitoring is especially crucial when sonic treatment or enzymatic digestion is part of the preparative procedure; an excessive sonication or contact with enzymes may alter spore surface detail.

FIXATION

The purpose of chemical fixation is to stabilize the microarchitecture of the cell. Ideally, fixation should preserve biological materials so that they closely resemble the living state. Some spores, particularly endospores of bacteria and ascospores of several Aspergilli, are sufficiently rigid to retain their shape during examination under high vacuum without any pretreatment. However, this characteristic is by no means true of all spores. Some spores may require chemical fixation to maintain natural resemblance even if sophisticated dehydration procedures such as critical point drying are used (Hayat and Zirkin, 1973; Cohen, in this volume).

When vegetative and spore-bearing structures are important, fixation is usually necessary to preserve morphological integrity. As noted in the previous section, many types of spore materials require extensive washing to provide a clean final preparation. Extremely fragile spores must be chemically fixed between preliminary washing with isotonic buffer and the final rinse in distilled water to avoid distortions caused by rapid osmotic shock. For less sensitive spore types, fixation is necessary only to preclude alteration of the natural shape during dehydration; judicious use of suitable fixatives can obviate the need for elaborate and lengthy dehydration procedures.

The fixative most widely used for spores is osmium tetroxide. Spores are fixed either by immersing in a 1% osmium tetroxide solution for 2 hr at room temperature (Williams, 1970) or by exposing to the vapors from a 1% solution for several hours (Locci *et al.*, 1971). The osmium solutions may be buffered depending upon the type of spore (Williams and Sharples, 1970). Although osmium tetroxide fixation followed by simple air-drying prevents or reduces collapse of some kinds of spores, this method is not adequate for many others. Certainly it is inefficient for vegetative structures.

These authors believe that aldehydes, such as formaldehyde and gluteraldehyde, are more effective than osmium tetroxide and are more satisfactory for highly collapsible spores. Chlamydospores of *Candida albicans* fixed for 1 hr in a 0.1% solution of gluteraldehyde in Krebs buffer, washed, and air-dried remain remarkably well preserved (Barnes *et al.*, 1971). Afrikian, St. Julian, and Bulla (unpublished data) find that vegetative cells of spore-forming bacteria are less subject to collapse after exposure to formalin vapors for 24 hr.

The following precautions are mentioned for the worker who is inexperienced in the general procedures of fixation: (1) After the specimens have been immersed in a fixative solution, they must be thoroughly rinsed with distilled water to remove salts from the cellular surfaces. (2) When the duration of fixation with gluteraldehyde is brief, a postfixation in osmium tetroxide may be required to stabilize osmotically sensitive spores during subsequent washings in distilled water; alternatively, longer fixation in the aldehyde may suffice. (3) Selection of a buffer will be dependent upon the fixative used as well as the molar strength of the fixative.

Routinely used buffers are veronal-acetate, s-Collidine, cacodylate, and phosphate (Hayat, 1972). Each one has advantages for different types of material; however, no single buffering system is universally superior. An extensive general coverage of fixation has been presented by Hayat (1970). In addition, the investigator confronted by a special fixative problem would be well advised to consult the transmission electron microscopy literature to ascertain the fixative-buffer solution(s) best suited for a given specimen.

DEHYDRATION

Spores should be in a highly dehydrated state for examination with the SEM. Water is an important structural component of living organisms, and its removal must be accomplished with extreme care in order to assure minimum damage to the specimen. The walls of many types of spores are so rigid that little obvious deformation results from simple air-drying. Indeed, air-drying is the most common method to dehydrate thick-walled spores of bacteria and some fungi.

After washing, the spores are placed either on glass squares cut from microscope slides, Millipore filters, or on pieces of filter paper and then exposed to atmospheric air at room temperature or to the warm breeze from a hair dryer. However, spores of many microorganisms collapse when dried in this manner. The extent to which such collapse hinders an investigation is relative.

Some of the deformation caused by air-drying can be avoided by an intervening dehydration in acetone or alcohol. Passage of wet spores through a graded series of solutions containing progressively increasing concentrations of the dehydrating agent and decreasing concentrations of water is generally suitable (Williams and Sharples, 1970; Baldacci et al., 1971; Fujita et al., 1971). The duration a sample must remain in each concentration is ~10 min, although large specimens require somewhat longer times.

The final stage of dehydration must be carried out carefully, and at least three changes of 100% acetone or ethanol are recommended. After the final change, the specimens are placed on a filter paper or glass squares and air-dried. A light microscope should be used to check that the ethanol or acetone dehydration procedure has not caused any serious alteration of the sample: ethanol induces mass separation of spores in chains with *Streptomyces griseus*, and causes the hairs on the spores of *S. finlayi* to become brittle and detach near their bases (Williams and Sharples, 1970).

More sophisticated means to circumvent problems of specimen damage associated with dehydration include freeze-drying (Rebhun, 1972), examination of specimens at low temperatures within the SEM, and critical point drying. Most simply stated, freeze-drying involves the rapid freezing of a specimen and subsequent removal of water by sublimation under high vacuum. As outlined for protozoa (Small and Marszalek, 1969), specimens are thoroughly washed in distilled water, placed onto the supercooled surface of an aluminum planchet floating on liquid nitrogen, and the resulting frozen microdrops transferred to a supercooled small-diameter aluminum foil disk which, in turn, is quickly positioned onto a precooled stage within a Pearse tissue dryer. The samples are sublimated at −60°C until completely dry, and then brought to room temperature before removal from the freeze dryer.

It must be stressed that at the time of quick-freezing, the specimen surface has to be free of fixative or buffering salts that would remain to mask the surface after sublimation. To minimize ice crystal damage which may occur during freeze-drying, physiological amounts of cryoprotective compounds such as glycerol, dimethyl sulfoxide, or polyvinyl pyrrolidone can be used; these compounds substantially lower the rate of ice crystal formation. For further information on their efficacy and the problems associated with their use, the reader is referred to Mac-Kenzie (1972) and to Nei, in this volume.

For low-temperature observation within the microscope, prefrozen samples are placed on a specimen stage cooled with circulating liquid nitrogen (Echlin *et al.*, 1970). When the temperature of the stage is raised from −100 to −90°C, surface ice quickly sublimes from the specimen owing to the vacuum of the microscope. Advantages of this technique are: minimum pretreatment of the specimen, and rapidity with which specimens can be examined after removal from their living environment. For the critical point drying procedure used to prepare microorganisms, the reader is referred to Cohen in this volume.

SPECIMEN MOUNTING

After spores have been dehydrated (if this step is part of the preparatory procedure), they are usually placed on a specimen stub before coating. It is rather easy to mount specimens that have been processed on platforms such as coverslips, glass squares cut from microscope slides, Millipore or Nuclepore filters, aluminum foil, and the like. The platforms are secured to the stub with double-coated adhesive tape, glue, paste, or similar adhesives. Spores capable of withstanding air-drying may be mounted directly on the metal specimen stub. When the latter method is used, natural adherence binds the spores firmly in place.

It is better to mount spores on glass instead of directly on the metal stubs for two reasons: (1) Glass mounts can be monitored easily in the light microscope before viewing with the SEM. (2) Glass provides a smooth electronically dark background in the final micrograph.

For relatively large specimens such as parasitized plant parts, it is generally necessary to use an adhesive to secure the specimen to the stub. The selection of a satisfactory adhesive is a prerequisite to satisfactory microscopic examination. Double-coated cellophane tape is used frequently, but there are a number of problems associated with its use. During examination under high vacuum, the adhesive may rise up around the lower portion of a spore resting directly on the tape, obscuring it from view. Coating a stub with the adhesive from cellophane tape dissolved in chloroform (1 cm/5 ml) may be more satisfactory, because a very thin layer of this material is far less likely to rise up and contaminate the spore.

Johari and DeNee (1972) compared a variety of adhesives, and recommend the use of organic glues such as Duco cement diluted with acetone, spray lacquers, and Mayer's albumen fixative. Regardless of the type of adhesive used, it must be allowed to dry to a tacky condition before the spores are mounted; otherwise they may sink into the adhesive material. During high-vacuum examination, any adhesive is a potential source of severe column and/or specimen contamination; thus only the smallest amount required to bind the specimen firmly to the stub should be used.

SPECIMEN SURFACE COATING

It is necessary to coat spores with a thin (10 to 15 nm) layer of conducting material prior to their examination with the microscope, because they, like other biological materials, are poor electrical conductors. Such coating serves to increase conductivity, to prevent buildup of a negative charge on the specimen surface, and to provide for the transference of excess thermal energy. Failure to accomplish either of the latter two objectives may result in serious distortion of the final image. The conducting substance is usually evaporated carbon and/or a metal such as aluminum, silver, gold, or palladium. A popular and efficient coating material is gold-palladium alloy.

The coating process is carried out in a high-vacuum evaporator where metal or carbon is vaporized onto the specimen. Ideally, the coating film should be uniformly continuous and of a thickness of about 10 nm. Practically, this thickness is difficult to achieve on spores with highly ornate surfaces; therefore the vacuum evaporator should possess a rotary shadower with a tilting device to facilitate the process. Such a device enables all areas of the specimen surface to be exposed to the vaporized material. When no tilting device is available, uniform coating is achieved by vaporizing the coating material from two separate positions, one at about a 30° angle to the center of the rotary plate and the other directly above the plate.

In some cases, it may be impractical or undesirable to apply a surface coating to the specimen. In such a case, the specimen must be observed at a considerably reduced accelerating voltage in order to avoid harmful charging effects on its surface. This procedure does not allow maximum resolution because of limited primary electron penetration of the specimen.

SPECIMEN STORAGE

Prolonged storage can adversely affect the quality of a specimen. Spores may prove highly satisfactory on examination soon after their preparation, and yet on reexamination several weeks later be utterly worthless. It is therefore recommended that the specimens be examined as soon as possible after vacuum-coating is completed. If a delay is unavoidable, the coated stubs should be stored in a closed container with a drying agent or under vacuum. In some cases it may be desirable to freeze the stub, but if this is done the stub must be allowed to return to room temperature before the container is opened to the air to prevent the condensation of atmospheric moisture on the cold stub.

APPLICATIONS

The ultimate test of a spore preparation is the scanning electron micrograph itself. In this section, we will illustrate the results expected with various types of spores viewed with the SEM.

Ascospores of Aspergilli

Some of the most beautiful ascospores of any of the fungi are those formed by members of the genus *Aspergillus*. These spores vary widely in form and ornamentation among the different species, and serve well to illustrate the diversity of the ascosporogenous fungi. In *Aspergillus fischeri* var. *glaber* (Fig. 7-1a), the lenticular ascospores have finely verrucose, convex surfaces. Each spore has two prominent equatorial crests that are irregularly polygonal; the surface of the crests is smooth.

In *A. quadricinctus* (Fig. 7-1b), the convex surfaces of the ascospores are more coarsely roughened, often with the formation of more or less pronounced short ridges and knobs. In this species, there are four rough crests, the two equatorial ones being the more prominent. According to Raper and Fennell (1965), *A. fischeri* var. *glaber* and *A. quadricinctus* are placed together in the *A. fumigatus* group. The ascospores of *A. cremeus* (in the *A. cremeus* group) also have two well-developed crests (Fig. 7-1c). Parallel to these crests there is an interrupted circular ridge. The central portion of the otherwise smooth, convex surfaces of the spores is ornamented by a number of short, irregularly distributed ridges.

In *A. spinulosus*, a member of the *A. ornatus* group, the ascospores are distinctly different from the other types presented. These spores lack

Fig. 7–1. Ascospores of Aspergilli.

(A) *Aspergillus fischeri var. glaber*
(B) *Aspergillus quadricinctus*
(C) *Aspergillus cremeus*
(D) *Aspergillus spinulosus*

any suggestion of equatorial crests; instead, they are densely covered by long spiny projections (Fig. 7–1d).

The ascospore morphology of four members of the *A. nidulans* group is presented in Fig. 7–2. Most spectacular are the delicate star-shaped ascospores of *A. variecolor* (Fig. 7–2a). Their unusual shape arises from the stellate outline of the two prominent equatorial crests. In *A. rugu-*

Fig. 7–2. Ascospores of Aspergilli (*Aspergillus nidulans* group).

(A) *Aspergillus variecolor*
(B) *Aspergillus rugulosus*
(C) *Aspergillus striatus*
(D) *Aspergillus nidulans*

losus (Fig. 7–2b), the convex surfaces of the spores are covered by thick, randomly branched ridges and small, round protuberances. The crests of these ascospores are smaller and marked by well-defined radial ribbing.

In contrast, the ascospores of *A. striatus* (Fig. 7–2c) which lack crests are ornamented by a series of thick ridges in the equatorial region. These ridges extend over the entire surface of the spores in a pattern that is often subparallel. In *A. nidulans,* the equatorial crests are radially ribbed (Fig. 7–2d) in much the same pattern that is evident in the crests of *A. rugulosus*, but the convex surfaces of the spores of *A. nidulans* are smooth and lack any ornamentation.

Yeast Ascospores

The morphology of ascospores is important in yeast classification; much of their surface detail is unresolved by light microscopy. The prominent equatorial ledge on the ascospores of *Schwanniomyces alluvius* (Fig. 7–3a) is apparent under the light microscope, but definition of any other distinguishing features is difficult. By scanning microscopy, it is possible to ascertain that the rough appearance observed with the light microscope is due to the presence of numerous fingerlike surface projections. These protuberances are slightly tapered with gently rounded tips.

By contrast, the surface roughening of the spores of *Debaryomyces hansenii* (Fig. 7–3b) results from a series of short, blunt ridges that are randomly distributed over the entire surface. *Saccharomycopsis vini* has unusual ascospores with two ledges and numerous warty outgrowths on the surface (Fig. 7–3c). The spores in this figure were air-dried, and show some signs of collapse. Such collapse can be prevented by critical point drying with the carbon dioxide-amyl acetate system. The equatorial ledge of the ascospores of *Pichia sargentensis* (Fig. 7–3d) is easily seen in the light microscope; the scanning electron microscope confirms that these spores are devoid of other surface ornamentation.

Spores of Fleshy Fungi

As discussed earlier, spores of fleshy fungi may be collected by the simple expedient of their discharge directly onto a suitably placed stub. The scanning electron micrographs presented here were prepared in this manner. Differences in resistance to collapse are well demonstrated by the three types of basidiospores shown in Fig. 7–4. The globose, spiny basidiospores of *Laccaria ochropurpurea* and *L. laccata* are highly resistant to collapse (Fig. 7–4a and b). In the latter micrograph, where the hilar appendage which attaches the spore to the basidium is shown, this appendage has retained its natural appearance.

In contrast, the basidiospores of *Agaricus silvicola* (Fig. 7–4c) tend to collapse when prepared in an identical manner. For the purpose of illustrating the smooth surfaces of these ellipsoid-oblong spores, such preparations are more than adequate. Ideally, however, one would prefer preparations that reflect in every aspect the natural conformation of the spores. Fig. 7–4d shows the lightly roughened surface of a *Morchella* sp. ascospore.

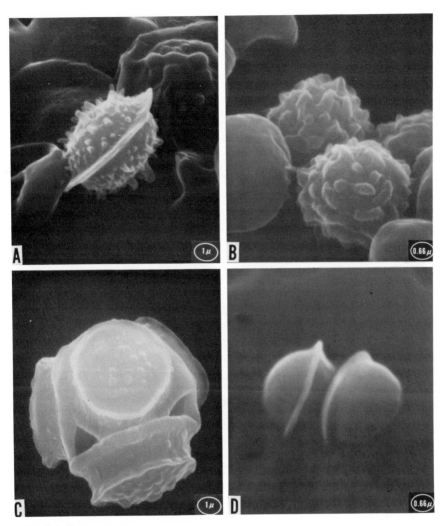

Fig. 7–3. Yeast ascospores.

    (A)   *Schwanniimyces alluvius*
    (B)   *Debaryomyces hansenii*
    (C)   *Saccharomycopsis vini*
    (D)   *Pichia sargentensis*

Bacterial Spores

Phase-contrast microscopy is generally unsatisfactory in resolving the
surface features of bacterial spores because of their small size. Under
phase-contrast optics, nearly all bacterial spores appear as smooth,

Fig. 7–4. Spores of fleshy fungi.

    (A)   *Laccaria ochropurpurea*
    (B)   *Laccaria laccata*
    (C)   *Agaricus silvicola*
    (D)   *Morchella* sp.

ellipsoidal or round bodies without any distinguishing surface features. The scanning microscope allows more accurate description. An intact sporulated cell of *Bacillus thuringiensis* (Fig. 7–5a) contains a parasporal body as well as a spore. Although the parasporal body (lower left) can be distinguished under phase contrast, its rhomboidal shape is apparent only under the electron microscope. The sporangium appears

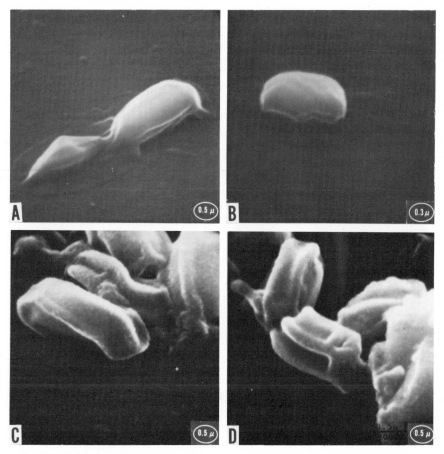

Fig. 7–5. Bacterial spores.

    (A)   *Bacillus thuringiensis*
    (B)   *Bacillus alvei*
    (C)   *Bacillus popilliae*
    (D)   *Bacillus lentimorbus*

in this figure as a transparent sheath draped over the spore and parasporal body forming a loose envelope.

An ungerminated spore of *B. alvei* (Fig. 7–5b) is shown within the intact sporangium. In this species, no parasporal bodies are formed. Free spores of *B. popilliae* (Fig. 7–5c) and *B. lentimorbus* (Fig. 7–5d) possess discrete, pronounced surface ridges that transverse the length of the spores. These ridges interconnect by short ridges on *B. lentimorbus* but not on *B. popilliae*.

Spores of Myxomycetes

The myxomycetes, or true slime molds, are classified by the way they form spores, the color of the spores, the shape of the mature fructification, and the amount of lime contained in the fructification. The presence and type of capillitium are also important characteristics in their

Fig. 7–6. Spores of Myxomycetes. Photographs courtesy of J. A. Murphy and L. L. Campbell.

 (A) *Lycogala epidendrum*

 (B) *Hemitrichia vesparium*

 (C) *Physarella oblonga*

 (D) *Lamproderma scintillans*

classification. *Lycogala epidendrum* is a slime mold with an aethaloid fructification. The spores of this species (Fig. 7–6a) are covered by delicate, reticulating ridges. *Hemitrichia vesparium* is characterized by the highly ornate sculpturing of its capillitial threads. The conspicuous spiral banding of these threads is apparent in Fig. 7–6b. Note especially the numerous spines that project from the banding ridges. The spores themselves are covered by small warts.

In *Physarella oblonga* (Fig. 7–6c), the capillitial threads are simple, slender filaments with occasional spinelike processes. The spores (collapsed in this preparation) have been described by light microscopy as nearly smooth, but under the scanning microscope they appear to have obvious warts. The capillitial threads of *Lamproderma scintillans* are straight, sparingly branched, and devoid of any conspicuous ornamentation (Fig. 7–6d). The spores which appear globose in uncollapsed preparations are roughly textured with warty protuberances arranged in nearly regular rows.

REFERENCES

Baldacci, E., Locci, R., and Baldan, B. P. (1971). On the spore formation process in Actinomycetes. III. Sporulation in *Streptomyces* species with hairy spore surface as detected by scanning electron microscopy. *Rivista di Patologia Vegetale*, Ser. IV, 7 suppl., 45.

Barnes, G., and Neve, N. F. B. (1968). Examination of plant surface microflora by the scanning electron microscope. *Trans. Brit. Mycol. Soc.* 51, 811.

Barnes, W. G., Flesher, A., Berger, A. E., and Arnold, J. D. (1971). Scanning electron microscopic studies of *Candida albicans*. *J. Bact.* 106, 276.

Bland, C. E., and Charles, T. M. (1972). Fine structure of *Pilobolus:* surface and wall structure. *Mycologia* 64, 774.

Bulla, L. A., Jr., St. Julian, G., Hesseltine, C. W., and Baker, F. L. (1973). Scanning electron microscopy. In: *Methods on Microbiology*, Vol. 8 (Norris, J. R., and Ribbons, D. W., eds.), p. 1. Academic Press, New York.

Bulla, L. A., Jr., St. Julian, G., Rhodes, R. A., and Hesseltine, C. W. (1969). Scanning electron and phase-contrast microscopy of bacterial spores. *Appl. Microbiol.* 18, 490.

Cole, G. T., and Aldrich, H. C. (1971). Ultrastructure of conidiogenesis in *Scopulariopsis brevicaulis*. *Can. J. Bot.* 49, 745.

Echlin, P., Paden, D., Dronzek, B., and Wayte, R. (1970). Scanning electron microscopy of labile biological material maintained under controlled conditions. In: *Scanning Electron Microscopy. Proc. 3rd Ann. Scan. Electron Micros. Symp.*, p. 273. I.I.T. Research Institute, Chicago.

Ellis, J. J., Bulla, L. A., Jr., St. Julian, G., and Hesseltine, C. W. (1971). Scanning electron microscopy of fungal and bacterial spores. In: *Scanning Elec-*

tron Microscopy. Proc. 4th Ann. Scan. Electron Micros. Symp., p. 145. I.I.T. Research Institute, Chicago.

Fujita, T., Tokunaga, J., and Inoue, H. (1971). *Atlas of Scanning Electron Microscopy in Medicine.* Igaku Shoin Ltd., Tokyo.

Greenwood, D., and O'Grady, F. (1969). Antibiotic-induced surface changes in microorganisms demonstrated by scanning electron microscopy. *Science* **163**, 1076.

Hawker, L. E. (1968). Wall ornamentation of ascospores of species of *Elaphomyces* as shown by the scanning electron microscope. *Trans. Brit. Mycol. Soc.* **51**, 493.

Hawker, L. E., and Gooday, M. A. (1968). Development of the zygospore wall in *Zygopus sexualis. J. Gen. Microbiol.* **54**, 13.

Hayat, M. A. (1970). *Principles and Techniques of Electron Microscopy: Biological Applications*, Vol. 1. Van Nostrand Reinhold Company, New York and London.

Hayat, M. A. (1972). *Basic Electron Microscopy Techniques.* Van Nostrand Reinhold Company, New York and London.

Hayat, M. A., and Zirkin, B. R. (1973). Critical point drying. In: *Principles and Techniques of Electron Microscopy: Biological Applications*, Vol. 3 (Hayat, M. A., ed.). Van Nostrand Reinhold Company, New York and London.

Johari, O., and DeNee, P. B. (1972). Handling, mounting and examination of particles for scanning electron microscopy. In: *Scanning Electron Microscopy. Proc. 5th Ann. Scan. Electron Micros. Symp.*, p. 249. I.I.T. Research Institute, Chicago.

Klainer, A. S., and Perkins, R. L. (1971). Surface effects of cell wall active antimicrobial agents. In: *Scanning Electron Microscopy. Proc. 4th Ann. Scan. Electron Micros. Symp.*, p. 329. I.I.T. Research Institute, Chicago.

Kurtzman, C. P., Smiley, M. J., and Baker, F. L. (1972). Scanning electron microscopy of ascospores of *Schwanniomyces. J. Bact.* **112**, 1380.

Leben, C. (1969). Colonization of soybean buds by bacteria: Observation with the scanning electron microscope. *Can. J. Microbiol.* **15**, 319.

Lewis, B. G., and Day, J. R. (1972). Behavior of uredospore germ-tubes of *Puccinia graminis tritici* in relation to the fine structure of wheat leaf surfaces. *Trans. Brit. Mycol. Soc.* **58**, 139.

Littlefield, L. J., and Bracker, C. E. (1971). Ultrastructure and development of urediospore ornamentation in *Melampsora lini. Can. J. Bot.* **49**, 2067.

Locci, R. (1969a). Direct observation by scanning electron microscopy of the invasion of grapevine leaf tissues by *Plasmopara viticola. Rivista di Patologia Vegetale*, Ser. IV, **5**, 199.

Locci, R. (1969b). Preliminary examination of microorganisms in soil by scanning electron microscopy. *Rivista di Patologia Vegetale*, Ser. IV, **5**, 167.

Locci, R. (1969c). Scanning electron microscopy of *Helminthosporium oryzae* on *Oryza sativa. Rivista di Patologia Vegetale*, Ser. IV. **5**, 179.

Locci, R. (1971). Scanning electron microscopy as an observation technique of biodeteriorated material microflora. *Rivista di Patologia Vegetale*, Ser. IV, **7**, 31.

Locci, R., and Bisiach, M. (1970). The *Phaseolus vulgaris—Uromyces appendiculatus* complex. I. Examination of the uredospore infection process by scanning electron microscopy. *Rivista di Patologia Vegetale*, Ser. IV, 6, 21.

Locci, R., and Bisiach, M. (1971). Scanning electron microscopy of the invasion of leaf tissues by the apple scab fungus. *Rivista di Patologia Vegetale*, Ser. IV, 7, 15.

Locci, R., Ferrante, G. M., and Rodrigues, C. J. (1971). Studies by transmission and scanning electron microscopy on the *Hemileia vastatrix—Verticillium hemileiae* association. *Rivista di Patologia Vegetale*, Ser. IV, 7, 127.

MacKenzie, A. P. (1972). Freezing, freeze-drying, and freeze-substitution. In: *Scanning Electron Microscopy. Proc. 5th Ann. Scan. Electron Micros. Symp.*, p. 273. I.I.T. Research Institute, Chicago.

McNeil, K. E., and Skerman, V. B. D. (1972). Examination of *Myxobacteria* by scanning electron microscopy. *Intern. J. Syst. Bacteriol.* 22, 243.

Old, K. M., and Robertson, W. M. (1969). Examination of conidia of *Cochliobolus sativus* recovered from natural soil using transmission and scanning electron microscopy. *Trans. Brit. Mycol. Soc.* 53, 217.

Raper, K. B., and Fennell, D. I. (1965). *The Genus Aspergillus*. The Williams & Wilkins Co., Baltimore.

Rebhun, L. I. (1972). Freeze-substitution and freeze-drying. In: *Principles and Techniques of Electron Microscopy: Biological Applications*, Vol. 2 (Hayat, M. A., ed.). Van Nostrand and Reinhold Company, New York and London.

Roth, I. L. (1971). Scanning electron microscopy of bacterial colonies. In: *Scanning Electron Microscopy. Proc. 3rd Ann. Scan. Electron Micros. Symp.*, p. 321. I.I.T. Research Institute, Chicago.

Rousseau, P., Halvorson, H. O., Bulla, L. A., Jr., and St. Julian, G. (1972). Germination and outgrowth of single spores of *Saccharomyces cerevisiae* viewed by scanning electron and phase-contrast microscopy. *J. Bact.* 109, 1232.

Sacks, L. E., and Alderton, G. (1961). Behavior of bacterial spores in aqueous polymer two-phase systems. *J. Bact.* 82, 331.

St. Julian, G., Bulla, L. A., Jr., and Hesseltine, C. W. (1971). Germination and outgrowth of *Bacillus thuringiensis* and *Bacillus alvei* spores viewed by scanning electron and phase-contrast microscopy. *Can. J. Microbiol.* 17, 373.

Small, E. B., and Marszalek, D. S. (1969). Scanning electron microscopy of fixed, frozen, and dried protozoa. *Science* 163, 1064.

Sullivan, J. L., Wagner, P. C., and DeBusk, A. G. (1972). Some observations of ascospores of *Neurospora crassa* made with a scanning electron microscope. *J. Bact.* 111, 825.

Watanakunakorn, C., Fass, R. J., Klainer, A. S., and Hamburger, M. (1971). Light and scanning-beam electron microscopy of wall-defective *Staphylococcus aureus* induced by lysostaphin. *Infec. Immunity* 4, 73.

Wickerham, L. J., and Kurtzman, C. P. (1971). Two new saturn-spored species of *Pichia. Mycologia* 63, 1013.

Williams, S. T. (1970). Further investigations of Actinomycetes by scanning electron microscopy. *J. Gen. Microbiol.* 62, 67.

Williams, S. T., and Davies, F. L. (1967). Use of a scanning electron microscope for the examination of Actinomycetes. *J. Gen. Microbiol.* **48**, 171.

Williams, S. T., and Sharples, G. P. (1970). A comparative study of spore formation in two *Streptomyces* species. *Microbios* **5**, 17.

8. THE AERIAL SURFACES OF HIGHER PLANTS

P. J. Holloway and E. A. Baker

Long Ashton Research Station, University of Bristol, Bristol, U.K.

INTRODUCTION

Higher plants are bounded either by a continuous layer of epidermal cells covered with a nonliving cuticular membrane or by a layer of cork cells. The cork layer (periderm) replaces the epidermis of organs which increase in thickness by secondary growth. In addition to basic anatomical and structural considerations, knowledge of the physical and chemical nature of plant surface structures is important in many fields of biological research, because a number of important physiological processes and other phenomena occur at these sites. Thus the surface layer plays a vital role in gaseous exchange and in the conservation of water in the plant, and prevents the loss of plant components by leaching. It affords protection against injuries due to wind, physical abrasion, frost, and radiation, and provides the first potential barrier against attack by fungi, insects, and other organisms.

The nature of the surfaces also greatly influences the deposition, retention, and distribution of chemicals (pesticides, growth regulators, and foliar nutrients) on plants. A great diversity of form and composition is shown by the surfaces of plants of different species. Some plants are strongly water-repellent, while others are easily wetted; water-repellency is greatest on plants which possess microcrystalline wax deposits or have a closed pattern of trichomes. The structure of the plant surface may also be modified by environmental conditions, microorganisms, pollution, and applied pesticides (Hull, 1970; Martin and Juniper, 1970).

Before 1967, information on the structure of plant surfaces was obtained with the aid of the light microscope or the transmission electron microscope (TEM). Plant surfaces were shown to have a complex topography extending from the macroscopic to the submicroscopic levels. The main disadvantage in the use of the light microscope is its restricted depth of focus rather than the inherent limits of magnification and resolution; however, it does permit the observation of the surface in its natural hydrated state. Because plant surfaces are generally dull and reflect little light, transparent replicas, in which some surface detail may be altered, are frequently used.

The TEM, with its high resolution, can be used to obtain information on surface structure only after preparing a suitable thin replica, which is a difficult and time-consuming procedure. Such replica methods often fail to provide reliable information on surfaces which have many re-entrant angles or large elevations and depressions. Too much emphasis has been placed in the past on carbon replicas examined at high magnifications with the TEM. The micrographs obtained can give false impressions of the microtopography of plant surfaces.

Scanning electron microscopy was first used in the study of plant surfaces by Amelunxen *et al.* (1967) and Holloway (1967). It was immediately recognized to be a valuable and versatile technique bridging the gap between the light and transmission electron microscopes with regard to resolution and magnification. For plant surface work, however, the scanning electron microscope (SEM) is best regarded as an extension of the range and quality of the light microscope by virtue of the excellent depth of focus which provides an easily comprehensible quasi-three-dimensional image of solid objects. The ability to examine the specimen at any angle in the electron beam enhances the three-dimensional effect.

The depth of focus available makes the SEM suitable even for applications in which the magnification and resolution of the light microscope are adequate. The most useful magnification range is between 50 and 15,000×, and the resolution varies between 20 and 100 nm according to accelerating voltage. Specimen preparation is relatively easy, and replication is usually not necessary because the thickness of the specimen is not important; the surface is observed directly. A wide range of plant surfaces have now been examined with the SEM.

In this chapter, the various techniques and problems associated with the SEM examination of the delicate surfaces of leaves, stems, and soft fruits will be described. Also included are methods suitable for studying isolated cuticular membranes and cork layers. Techniques for more robust plant specimens such as pollen grains, seeds, and hard fruits are dealt with by other authors in this volume and by Burrichter *et al.* (1968), Echlin (1968), Reyre (1968), and Heywood (1969). The micrographs presented here were obtained using a Cambridge Stereoscan Mark 2A scanning electron microscope; the operating conditions described throughout also apply to this instrument.

SPECIMEN PREPARATION

Fresh Materials

Pieces of tissue (up to 0.5 cm^2) are carefully detached from the plant and mounted directly on standard aluminum stubs. Extreme care is necessary in handling specimens, especially for examining wax structures because they are easily damaged by any physical contact. The intact leaf tissue of most plants can be used; but for xerophytic leaves, fruits, and stems, it is advisable to remove the surface layers with the minimum of adhering tissue by slicing through the epidermal cells just below the cuticle. This procedure effectively reduces the moisture content of the specimen and minimizes the risk of distortion during drying (Heslop-Harrison and Heslop-Harrison, 1969; Baker and Parsons, 1971). The adaxial and abaxial surfaces of leaf specimens can be mounted on the same stub to facilitate comparison in the microscope.

A variety of proprietary adhesives can be used to attach the specimen to the stub (Muir and Rampley, 1969); in our experience, double-sided adhesive tape is most suitable for plant surface work. However, the adhesion of moist specimens (tissue slices) to adhesive tape is poor, and a solvent-based glue (Durofix) is recommended. When using a glue, the solvents must be allowed to partially evaporate before mounting the specimen in order to minimize any interaction with surface structures, especially wax.

For the examination of waxy specimens, additional procedures are used. Removal of wax may be necessary to allow an examination of stomata or other surface features obscured by the wax layer. It is accomplished by washing the surface briefly with petrol/acetone (4:1) (Amelunxen *et al.*, 1967), chloroform (Holloway, 1967; Baker and Holloway, 1971), xylene, or acetone (Verdus, 1969). Washing with solvent can also be used as a routine procedure to confirm that wax is indeed present on the surface (Baker and Holloway, 1971); the trichomes of some plants closely resemble in appearance the wax structures found on others.

Useful information on the constitution of wax layers can be obtained by lightly abrading a small area of the specimen surface with a fine wire.

By comparison with the unabraded layer, this procedure enables an examination of the manner in which the wax layer is built up from the cuticular surface, and is especially useful where the wax layer has a composite structure (Baker and Holloway, 1971). These modified surfaces may conveniently be mounted with an intact specimen on the same stub to facilitate comparison.

Preserved Material

The preservation of plant surface material is usually not necessary for examination with the SEM except when the specimen cannot tolerate removal of water by the vacuum system of the coating unit or the microscope. Such specimens usually have a high moisture content or possess a very thin epidermis. Simple air-drying after mounting on the stub may be satisfactory in some cases (Campbell, 1972); however, it can produce some distortion of the epidermal cell surface (Heslop-Harrison and Heslop-Harrison, 1969). Chemical fixation of specimens has been used (Albrigo and Brown, 1970; Zachariah and Pasternak, 1970; Panessa and Gennaro, 1972), but was found to yield inferior results to those obtained from fresh, unfixed specimens (Falk et al., 1971). Critical point drying, using amyl acetate for dehydration, was used by Royle and Thomas (1971) for the study of stomatal development in hop leaves. The procedure, however, may not completely remove water, and may cause some bulk shrinkage or produce violent bubbling (Boyde and Wood, 1969; Boyde, 1972).

Several workers have employed freeze-drying or quench-freezing in liquid nitrogen or Freon (12 or 22) (Amelunxen et al., 1967; Einert et al., 1970; Campbell, 1972; Troughton and Donaldson, 1972). The major disadvantage is the possibility of ice crystal artifacts, although the development of a low-temperature stage for the SEM may alleviate this problem (Echlin et al., 1970). Preservation techniques involving fixation or the use of organic solvents (e.g., critical point drying) are not suitable for studying waxy specimens, because wax will be dissolved from the surface. Detailed accounts of dehydration methods suitable for labile specimens have been presented (Boyde and Wood, 1969; Boyde, 1971 and 1972; Echlin, 1971 and 1972).

Some workers have employed replica techniques, but these have only a limited application. Replication can yield false information on the morphology of stomata (Idle, 1969) and wax formations (Baker and Holloway, 1971), and fails to yield natural images of many trichomes. Chapman (1967) has used polystyrene negatives, while others have used a silicone rubber negative to prepare a positive in cellulose acetate (Idle,

1969) or resin (Sacalis, 1972; Schwerdtfeger, 1972). The SEM micrographs of apple fruit replicas obtained by Schwerdtfeger (1972) are in agreement with those obtained by direct examination of fresh apple surfaces after metal-coating (Fisher and Corke, 1971).

Isolated Cuticular Membranes and Cork Layers

Cuticular membranes are obtained from leaves and fruits by first removing small pieces of tissue; disks (1-2 cm²), cut with a cork borer, are easy to handle. Wax is removed by washing with chloroform and the membranes are detached from the tissue using a solution of ammonium oxalate (1.6%) and oxalic acid (0.4%) (Huelin and Gallop, 1951), pectic enzyme solution (5%) at pH 3.5 (Orgell, 1955) or zinc chloride-hydrochloric acid solution (1 gm $ZnCl_2$ in 1.7 ml concentrated HCl) (Holloway and Baker, 1968). If oxalate or pectinase is used, adhering cell fragments are removed from the isolated membranes by further treatment with cellulase or zinc chloride-hydrochloric acid. Membranes from more robust plants (gymnosperms) and fossilized leaves can be obtained after maceration of tissue with concentrated nitric acid or Schultze's reagent (Alvin, 1970; Boulter, 1970 and 1971). Isolated membranes are exhaustively extracted with organic solvents to remove wax and pigments and stored in water or aqueous methanol before mounting.

The methods used for mounting depend mainly on the thickness of the membrane and its susceptibility to shrinkage (Baker and Parsons, 1971). Robust specimens can be attached directly to stubs with a suitable glue (Alvin, 1970; Boulter, 1971). Thick membranes do not shrink when air-dried on a porous plate, and are mounted on stubs with double-sided adhesive tape. Less thick membranes which shrink when air-dried are floated from distilled water onto stubs covered with Scotchmount (Minnesota Mining and Manufacturing Company). Excess water is blotted from the stub without touching the membrane and then air-dried for 30 min.

An alternative procedure for thin membranes is to suspend them in methanol and then transfer them to distilled water, where they float fully expanded on the surface of the water. The membranes are removed from the water by placing a piece of filter paper underneath them and dried in a stream of nitrogen directed at the membrane through the filter paper. The dried membranes can then be lifted from the paper with fine tweezers and mounted. Extremely delicate membranes are best mounted directly onto stubs while floating on distilled water and then air-dried. A dissecting microscope is usually necessary when mounting these membranes.

Cork layers are obtained by stripping them from stems or natural exfoliations of some plants (e.g., silver birch). Specimens of suitable size are mounted directly onto stubs with adhesive tape. The cellular structure of the layers can be revealed by slicing with a sharp blade; cell inclusions and the wall structure can then also be studied (Jensen, 1972).

SPECIMEN DEHYDRATION AND COATING

The dehydration of fresh plant materials is accomplished, without appreciable shrinkage, by the vacuum system of the coating unit before coating with metal. It is advisable to reduce the pressure gradually over a period of 15 to 30 min (Baker and Parsons, 1971). Most plant cells have rigid cellulose walls that are further strengthened by such resistant materials as cutin, suberin, and lignin in the surface tissues. This strengthening enables them to resist collapse following removal of water. Damage from desiccation is most likely to occur in leaves which owe their rigidity mainly to the turgor of the epidermal, palisade, and mesophyll cells. Damage is least likely in a specimen where rigidity is provided by mechanical tissues or where water loss is restricted in any event by a thick cuticle or wax deposit (Heslop-Harrison and Heslop-Harrison, 1969).

The most common form of vacuum damage is the distortion or collapse of the surface of the epidermal cells; examples of this damage are shown on *Stachys sylvatica* (Fig. 8–1) and *Trifolium repens* (Fig. 8–5) leaves. The collapse of delicate trichomes—for example, in *Chenopodium album* and *T. repens* (Fig. 8–2)—is also occasionally troublesome. If vacuum damage is suspected, it can be vindicated in most cases by a comparison with the appearance of the natural surface using reflected light and the light microscope.

In order to obtain satisfactory images at high magnification and resolution, plant surfaces (or replicas thereof) should be rendered conducting. For this purpose a thin film of gold (Heslop-Harrison and Heslop-Harrison, 1969; Newton, 1972; Schwerdtfeger, 1972) or gold-palladium alloy (Au 60:Pd 40) (Holloway, 1967; Echlin, 1968; Heslop-Harrison and Heslop-Harrison, 1969; Baker, 1970; Still *et al.*, 1970; Baker and Holloway, 1971) is evaporated directly onto the surface.

A coating of carbon (Amelunxen *et al.*, 1967; Einert *et al.*, 1970) or an initial coating of carbon followed by one of gold has also been used (Amelunxen *et al.*, 1967; Johnson and Jeffree, 1970; Jeffree *et al.*, 1971; Panessa and Gennaro, 1972). A carbon coating is claimed to stabilize the surface and reduce cracking, and to provide a satisfactory substrate to which the gold layer is firmly adhered (Boyde and Wood, 1969; Panessa and Gennaro, 1972). Carbon is also easy to evaporate and scatter,

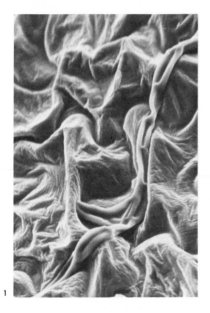 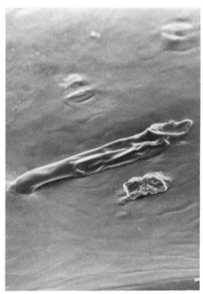

All specimens were coated with gold-palladium. Micrographs were recorded on Ilford HP4 film at 10 kV (except where otherwise stated) with the specimen at 45° to the electron beam.

Fig. 8–1. Vacuum damage. Contortion of epidermal cells on adaxial surface of *Stachys sylvatica* leaf. 1,300 X.

Fig. 8–2. Vacuum damage. Collapsed trichome on abaxial surface of *Trifolium repens* leaf. 550 X.

ensuring that a complete film is deposited particularly over rough surfaces (Amelunxen *et al.*, 1967; Boyde and Wood, 1969). The conducting materials are evaporated from a heated tungsten filament under vacuums of the order of 5×10^{-4} to 8×10^{-5} torr (Cross, 1972a).

The object of specimen coating is twofold: (1) to prevent the build up of electric charge on the surface, and (2) to increase the emission of low-energy secondary electrons that are responsible for contrast in the microscope image. It is important to ensure that all of the surface of even the roughest specimen of plant material is made conducting if artifacts due to charging are to be avoided. Adequate coating can be achieved only by rotating the specimen and at the same time tilting about two mutually perpendicular axes (Echlin, 1968; Boyde and Wood, 1969; Baker and Holloway, 1971). Care should also be taken that specimens are smaller than the stub, so that the conducting coat can continue unbroken from the specimen to the edge of the stub in order to provide a conducting path to the earth.

Most workers have not determined with any accuracy the thickness of the conducting layers applied, but a layer 30 to 50 nm thick is frequently quoted. As pointed out by Cross (1972a), such statements are misleading because the layer is inevitably nonuniform, especially over a rough surface, so that average thickness has little meaning. However, the use of new techniques for measuring film thickness may remove this source of error (Echlin and Hyde, 1972). In practice, the aim should be to apply the thinnest possible layer which will suppress charging, and it is best determined empirically for a given specimen. The thickness of the layer required will depend upon both the electrical properties of the specimen and its real surface area; the more complex the topography of the specimen, the more metal will have to be applied.

The intensity of the secondary electron emission from a specimen is governed chiefly by the atomic number of the target (review by Northrop, 1972). A satisfactory emission of secondary electrons is obtained from a gold coating, which is not bettered by any other practicable material (Echlin, 1968; Boyde and Wood, 1969); the emission from carbon films alone is poor (Amelunxen et al., 1967). Silver, however, is recommended where the ability to contour accurately is of prime consideration (Cross, 1972a), although the metal has not been widely used for plant specimens.

EXAMINATION WITH THE MICROSCOPE

Uncoated Specimens

In some instances, satisfactory images, with medium resolution, have been obtained from uncoated plant specimens placed directly in the vacuum system of the microscope (Heslop-Harrison and Heslop-Harrison, 1969; Heslop-Harrison, 1970; Falk et al., 1970 and 1971; Murr, 1970; Echlin and Hyde, 1972; see Howden and Ling, and Nei, in this volume). The pumping system can tolerate the water loss from small, partly hydrated leaf fragments, but radical structural changes caused by desiccation inevitably occur. Reliable information, however, can be obtained for periods up to 15 min. Dissolved ions and minerals in the cell sap of the specimen probably generate the secondary electrons necessary to form the images as well as to provide a suitable conductive pathway (Heslop-Harrison and Heslop-Harrison, 1969; Echlin, 1971).

In order to avoid adverse charging effects on the specimens, it is necessary to operate the instrument at low accelerating voltages, usually below 5 kV, to reduce beam penetration and the radiation flux impinging on the specimen. Resolutions of 80 to 90 nm are possible at voltages of

1 to 2 kV (Echlin, 1971). Although the technique is not widely used, it does enable an image to be obtained within a few minutes after the specimen has been removed from the living plant.

Coated Specimens

In order to use the SEM to its full potential of magnification and resolution, most investigations of plant surfaces are carried out using coated specimens. Examination with the microscope involves the selection of the optimum combination of instrument settings to obtain the best image of a given specimen or to yield the surface information required. The major factor governing the operation of the instrument is the interaction of the electron beam with the plant specimen. The most serious effects result from (1) penetration of the beam into the specimen, and (2) electron bombardment of the surface. The first effect causes a deterioration in the quality of the image; the second, irreversible damage to the specimen. It is often not possible to eliminate all detrimental effects, and compromises have to be made in settings of the instrument such as accelerating voltage, lens currents, and final objective aperture. The general guidelines described by Boyde and Wood (1969), Hearle *et al.* (1970), Baker and Holloway (1971), Baker and Parsons (1971) and Cross (1972b) should be followed.

The main problem with beam penetration is charging, since secondary electrons are emitted from a volume below the specimen surface. If penetration is severe, the whole specimen may charge (bulk charging); examples of this phenomenon are shown on the waxy surface of *Avena sativa* leaf (Fig. 8–3) and on the trichomes of *Senecio laxifolius* leaf (Fig. 8–4). These micrographs show extreme contrast between bright and dark areas. The charged areas have also deflected and distorted the beam, causing the banding effect on *A. sativa* and the distorted image of *S. laxifolius*. This results in an overall loss of resolution. Bulk charging effects also occur if specimens are not coated, if the metal coating is incomplete or too thin, or if the specimen is not adequately grounded to the stub.

Penetration of the beam, however, may be confined to particular features of the specimen surface, causing an edge contrast or spike brightness effect. Examples of this phenomenon are shown in Figs. 8–5 and 8–6; the outlines of the epidermal cells of *T. repens* and the ridges of the inner surface of the isolated cuticular membrane of *Agave americana* are excessively bright, with a consequent loss of structural detail. The effect is attributed to secondary electrons generated in depth from the surface, where it presents an edge or declivity which is approximately parallel

Fig. 8–3. Bulk charging. Wax deposits on abaxial surface of *Avena sativa* leaf. 1,250 X.
Fig. 8–4. Bulk charging. Trichomes on abaxial surface of *Senecio laxifolus* leaf. 1,350 X.

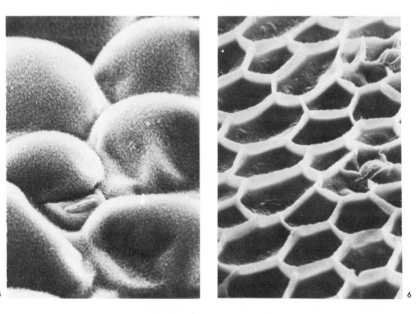

Fig. 8–5. Vacuum damage and beam penetration. Collapsed epidermal cells and edge contrast artifact on adaxial surface of *Trifolium repens* leaf, 15 kV. 1,420 X.
Fig. 8–6. Beam penetration. Edge contrast artifact on the ridges of the inner surface of the isolated cuticular membrane of *Agave americana*, 20 kV. 430 X.

to the direction of the beam. Penetration effectively produces a larger proportion of surface per image point (Boyde and Wood, 1969; Boyde, 1971).

More serious effects are produced by electron bombardment of the specimen, either directly or indirectly owing to the effect of heating. Beam damage is usually seen taking place during observation, and is caused by the dissipation of the beam energy over a smaller area as the magnification of the microscope is increased (the area scanned at a magnification of 10,000 is 10 μm^2). When the magnification is reduced, the damaged area of the surface can be seen to correspond with the raster used at the higher magnification.

The susceptibility of plant material to beam damage, as well as the nature of the damage which may arise, is variable. The wax deposits on most leaves will melt if adequate precautions are not taken, and damage may become apparent at magnifications as low as 2,000×. An example of the severe damage which can take place is shown on *Eucalyptus globulus* leaf (Fig. 8–7). However, there appears to be no definite relationship between the melting point of the surface wax and its susceptibility to beam damage; plants with delicate wax structures which project well from the surface are most susceptible (Baker and Holloway, 1971). The surfaces of some plant specimens, especially those of young leaves and isolated cuticular membranes, may blister or crack and be completely disrupted (Baker and Parsons, 1971; Davis, 1971). An example of this type of damage is shown on *Malus pumila* leaf (Fig. 8–8).

The occurrence of artifacts due to penetration and damage can be substantially reduced or eliminated by lowering the energy of the electron beam. Some plant materials can be examined at more than 20 kV but others may need 5 kV or less; resolutions of 40 to 50 nm are still attainable at voltages below 10 kV (Boyde and Wood, 1969; Hearle *et al.*, 1970). Any reduction of accelerating voltage, however, increases the diameter of the electron beam, which must be adjusted to a usable value by increasing the lens currents in order to maintain adequate resolution.

A further increase of the lens currents to reduce the intensity of the electron beam may also be used to minimize damage to the specimen. Thus a voltage of 10 kV and lens currents of the order of 0.7 to 0.9 Amps are necessary for the study of most waxy leaf surfaces (Baker and Holloway, 1971), but the same voltage and currents of 0.4 to 0.6 Amps are adequate for other plant surfaces (Baker and Parsons, 1971). The use of high lens currents invariably increases the signal-to-noise ratio. Therefore, to preserve the quality of the image, photographs may have to be taken at scan times of up to 100 sec.

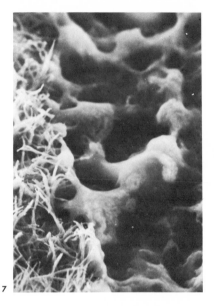
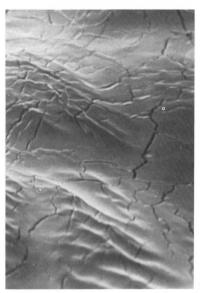

7 · 8

Fig. 8–7. Beam damage. Melted wax deposits on adaxial surface of *Eucalyptus globulus* leaf, 15 kV. 2,220 X.
Fig. 8–8. Beam damage. Cracking of adaxial surface of *Malus pumila* leaf, 15 kV. 2,300 X.

The best resolution and depth of focus at a high magnification are obtained using a short working distance and a large final objective aperture (200 μm), which yields a sharply defined electron beam. At lower magnifications, a smaller aperture together with a larger working distance may be necessary to obtain the maximum depth of focus. The lens currents can then be reduced to an extent depending upon the susceptibility of the specimen to beam damage (Boyde and Wood, 1969; Hearle *et al.*, 1970; Cross, 1972b).

Another important factor which influences the quality of the SEM image is the angle of incidence of the primary electron beam on the specimen surface. This angle governs the intensity of secondary electron emission; emission is lowest when the surface is normal to the beam, and increases as the surface is tilted through 90° to face the electron collector (review by Northrop, 1972). In practice, a suitable working compromise has to be found; for most specimens, a tilt angle of 45° is used. However, on plants which have trichomes or other extrusions which project perpendicularly from the surface, the specimen is best viewed at ~90° to the beam (Cross, 1972b).

Stereology

Stereopair images are necessary if precise measurements of heights and volumes or reconstruction of original three-dimensional morphology are required from SEM micrographs (review by Hilliard, 1972; Lane, 1972; Welford, 1972). There is usually considerable foreshortening in the image because of the angle of inclination of the specimen relative to the electron beam. True scale depth is obtained by stereoscopic examination of micrographs obtained at two different tilt angles of 8° and 10°.

APPLICATIONS

The SEM has so far been used chiefly to obtain information on plant surface morphology with a new perspective (Amelunxen et al., 1967; Chapman, 1967; Holloway, 1967 and 1971; Heslop-Harrison and Heslop-Harrison, 1969; Heywood, 1969; Idle, 1969; Lange, 1969; Verdus, 1969; Baker, 1970 and 1971; Johnson and Jeffree, 1970; Baker and Holloway, 1971; Baker and Parsons, 1971; Davis, 1971; Hanover and Reicosky, 1971; Jeffree et al., 1971; Rentschler, 1971; Albrigo, 1972; Jensen, 1972; Schwerdtfeger, 1972; Troughton and Donaldson, 1972).

However, the interactions of the plant surface with its environment (Murr, 1970; Baker, 1972), insects (Heslop-Harrison, 1970; Gibson, 1971), fungi (Locci, 1969 and 1971a and b; Fisher and Corke, 1971; Locci and Bisiach, 1971; Locci and Quaroni, 1971; Pugh and Buckley, 1971; Purnell, 1971; Royle and Thomas, 1971; Campbell, 1972), viruses (Conti and Locci, 1972), and agricultural chemicals (Albrigo and Brown, 1970; Still et al., 1970; Bukovac et al., 1971; Sacalis, 1972) have also been studied with the aid of the SEM. The technique is certain to find more uses by botanists and paleobotanists in taxonomic studies (Alvin, 1970; Boulter, 1970 and 1971; Verdus, 1970; Newton, 1972) and in the ontogeny of plant organs (Einert et al., 1970; Falk et al., 1970).

The wide application and versatility of the SEM for plant surface work is best illustrated by the series of micrographs (Figs. 8–9 through 8–32). The magnification facilities of the microscope enable a better understanding of the spatial relations of features of microtopography.

On *Fragaria chiloensis* leaf the lower magnifications (Figs. 8–9 and 8–10) reveal the patterns of venation and the distribution of trichomes, stomata and wax over the surface, whilst the higher magnifications (Figs. 8–11 and 8–12) show the fine structure of the glandular trichomes and

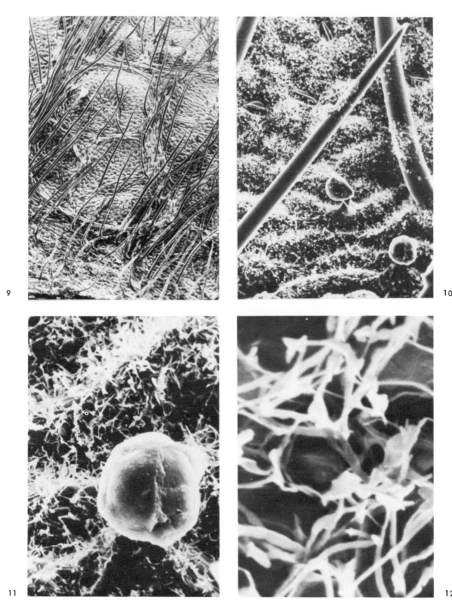

Fig. 8–9—8–12. Abaxial surface of *Fragaria chiloensis* leaf

Fig. 8–9. General view of surface. 50 X.

Fig. 8–10. Detail of covering trichomes, distribution of stomata, glandular trichomes, and wax. 500 X.

Fig. 8–11. Detail of glandular trichome and wax. 2,000 X.

Fig. 8–12. Ultrastructure of wax deposits. 10,500 X.

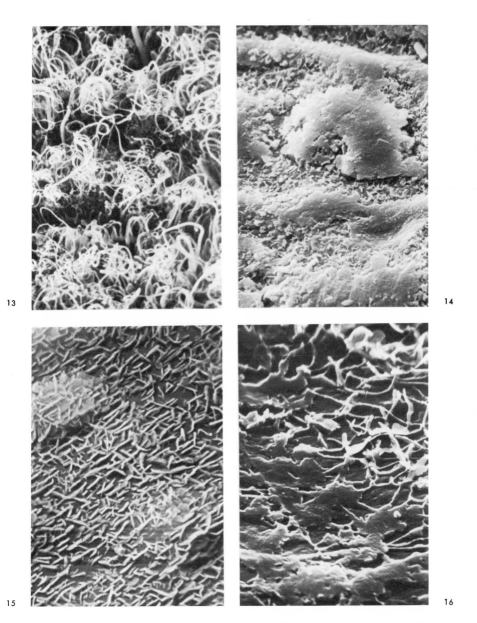

Fig. 8–13. Dense meshwork of wax threads on adaxial surface of *Musa sapientum* leaf. 800 X.

Fig. 8–14. Wax aggregates on adaxial surface of *Phragmites communis* leaf. 2,800 X.

Fig. 8–15. Acicular wax crystals on stem surface of *Equisetum arvense*, 25 kV. 10,500 X.

Fig. 8–16. Abaxial surface of Rose cv. "Bacarra" leaf. Unabraded surface (top right) covered by wax plates and ribbons, abraded surface (bottom left) showing the underlying wax crust, 15 kV. 5,400 X.

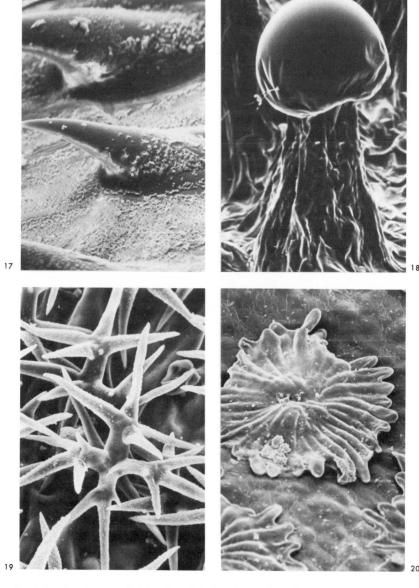

Fig. 8–17. Asperities on adaxial surface of *Arrhenatherum elatius* leaf. 1,325 X.

Fig. 8–18. Glandular trichome on adaxial surface of *Cannabis sativa* bracteole. 525 X.

Fig. 8–19. Dense meshwork of "candelabra" covering trichomes on adaxial surface of *Lavandula intermedia*. 450 X.

Fig. 8–20. Peltate covering trichomes on adaxial surface of *Olea europaea* leaf. 600 X.

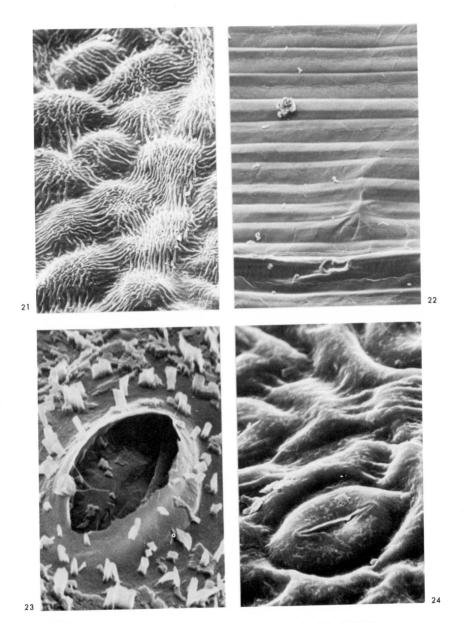

Fig. 8–21. Cuticular striations on adaxial surface of *Syringa vulgaris* leaf. 540 X.

Fig. 8–22. Smooth cuticle on adaxial surface of *Lolium multiflorum* leaf. 300 X.

Fig. 8–23. Distribution of wax deposits around stoma and within substomatal chamber on abaxial surface of *Brassica oleracea* var. *gemmifera*. 6,200 X.

Fig. 8–24. Detail of stoma and cuticular surface on abaxial surface of *Ligustrum vulgare*, 15 kV. 2,060 X.

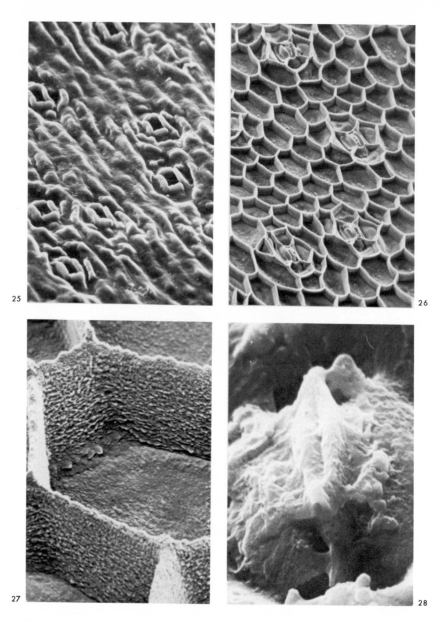

Fig. 8–25. Outer surface of abaxial cuticular membrane of *Agave americana*. 190 X.

Fig. 8–26. Inner surface of abaxial cuticular membrane of *Agave americana* leaf showing that the cuticle extends between the anticlinal walls of the epidermal cells. 220 X.

Fig. 8–27. Detail of inner surface of abaxial cuticular membrane of *Agave americana* leaf showing extensively pitted surface. 1,660 X.

Fig. 8–28. Detail of cuticle covering the stomatal apparatus on the inner surface of abaxial cuticular membrane of *Citrus aurantifolia*. 6,500 X.

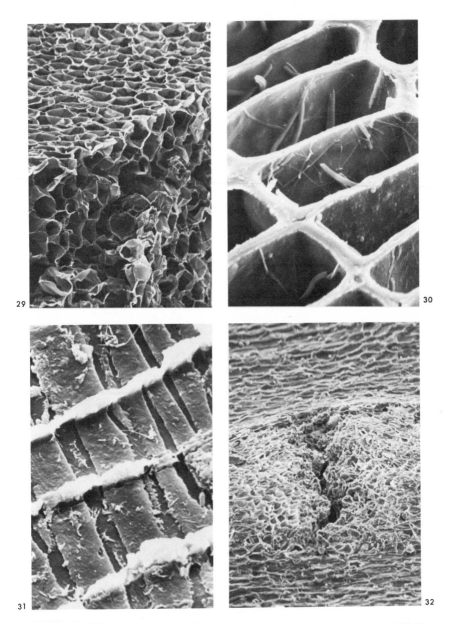

Fig. 8–29. Section of cork layer from *Quercus suber* stem showing cellular structure. 210 X.

Fig. 8–30. Detail of T.S. of cork layer from *Quercus suber* stem showing arrangement of individual cork cells and the crystalline triterpenoid cell inclusions. 1,800 X.

Fig. 8–31. Detail of inner surface of exfoliated cork layer of *Betula pendula* stem showing cellular structure. 2,600 X.

Fig. 8–32. Structure of lenticel on the stem surface of *Acer griseum*. 220 X.

wax deposits. The SEM is the only means of obtaining accurate information on the morphology of wax deposits, which are delicate, have a complex structure, or project vertically from the surface.

The carbon replica technique, used in conjunction with the transmission electron microscope, often damages and thereby modifies the morphology of these waxes, yielding false information. The transmission electron microscope, however, is usually required if information on the ultrastructure of individual wax particles is needed, because such detail falls outside the resolution of the SEM. Some examples of plant surface waxes are shown in Figs. 8–13, 8–14, and 8–15; a dense meshwork of threads on *Musa sapientum* leaf, aggregates on *Phragmites communis* leaf and small acicular crystals on the stem of *Equisetum arvense*. The use of the abrasion technique is illustrated on rose leaf (Fig. 8–16); the abraded surface reveals that the wax plates and ribbons visible on the surface are superimposed upon an amorphous wax crust which covers the whole cuticle surface.

The depth of focus provided by the SEM makes it valuable in the study of the structure and distribution of trichomes. Shown with the aid of the SEM are: the characteristic asperities found on grasses (Fig. 8–17), a single glandular trichome of *Cannabis sativa* (Fig. 8–18), a complex pattern of "candelabra" trichomes on *Lavandula intermedia* (Fig. 8–19) and the flattened peltate trichomes of *Olea europaea* (Fig. 8–20). Some other features of the plant surface revealed by the SEM are shown in Figs. 8–21, 8–22, 8–23, and 8–24. The cuticular surface may have ornamentation (e.g., *Syringa vulgaris*) (Fig. 8–21), or may be smooth (e.g., *Lolium multiflorum* (Fig. 8–22)). The amount of structural information that can be obtained by examining stomata is illustrated on *Brassica oleracea* (Fig. 8–23) and *Ligustrum vulgare* (Fig. 8–24) leaves. If the stoma is open, observation within the stomatal apparatus is possible.

Studies of the surface structure of isolated cuticular membranes using the SEM are important in revealing the degree of development (Fig. 8–25) and, in particular, the relationship between the membrane and the underlying epidermal cells (Figs. 8–26 and 8–27) and the extent of cutinization of the stomatal apparatus (Fig. 8–28) or other epidermal features. Examples of the use of the microscope in the examination of cork layers are shown in Figs. 8–29, 8–30, 8–31, and 8–32. The architecture of the layers and their cellular constitution can easily be demonstrated by simple sectioning (Figs. 8–29 and 8–30) and by observing the surface (Fig. 8–31). The structure of lenticles can also be studied (Fig. 8–32).

We are grateful to Mrs. E. Parsons, Mrs. B. Bole, Mr. D. J. Hall, and Mr. D. J. Alden for their skillful operation of the SEM and to Mrs. J. A. Sadd for the photographic work.

REFERENCES

Albrigo, L. G. (1972). Distribution of stomata and epicuticular waxes on oranges as related to stem and rind breakdown and water loss. *J. Amer. Hort. Sci.* **97,** 220.

Albrigo, L. G., and Brown, G. E. (1970). Orange peel topography as affected by a preharvest plastic spray. *Hort. Sci.* **5,** 470.

Alvin, K. L. (1970). The study of fossil leaves by scanning electron microscopy. In: *Proc. 3rd Ann. Scan. Electron Micros. Symp.,* p. 123. I.I.T. Research Institute, Chicago.

Amelunxen, F., Morgenroth, K., and Picksak, T. (1967). Untersuchungen an der Epidermis mit dem Stereoscan-Elektronenmikroskop. *Z. Pflanzenphysiol.* **57,** 79.

Baker, E. A. (1970). The morphology and composition of isolated plant cuticles. *New Phytol.* **69,** 1053.

Baker, E. A. (1971). Chemical and physical characteristics of cuticular membranes. In: *Ecology of Leaf Surface Micro-organisms* (Preece, T. F., and Dickinson, C. H., eds.), p. 55. Academic Press, London.

Baker, E. A. (1972). The effect of environmental factors on the development of the leaf wax of *Brassica oleracea* var. *gemmifera.* M.Sc. thesis, University of Bristol, U. K.

Baker, E. A., and Holloway, P. J. (1971). Scanning electron microscopy of waxes on plant surfaces. *Micron* **2,** 364.

Baker, E. A., and Parsons, E. (1971). Scanning electron microscopy of plant cuticles. *J. Microscopy* **94,** 39.

Boulter, M. C. (1970). Lignified guard cell thickenings in the leaves of some modern and fossil species of Taxodiaceae (Gymnospermae). *J. Linn. Soc. (Biol.)* **2,** 41.

Boulter, M. C. (1971). Fine details of some fossil and recent conifer leaf cuticles. In: *Scanning Electron Microscopy: Systematic and Evolutionary Applications.* The Systematics Association Special Volume 4 (Heywood, V. H., ed.), p. 211. Academic Press, London.

Boyde, A. (1971). A review of problems of interpretation of the scanning electron microscope image with special regard to methods of specimen preparation. In: *Proc. 4th Ann. Scan. Electron Micros. Symp.,* p. 1. I.I.T. Research Institute, Chicago.

Boyde, A. (1972). Biological specimen preparation for the scanning electron microscope: an overview. In: *Proc. 5th Ann. Scan. Electron Micros. Symp.,* p. 257. I.I.T. Research Institute, Chicago.

Boyde, A., and Wood, C. (1969). Preparation of animal tissues for surface-scanning electron microscopy. *J. Microscopy* **90,** 221.

Bukovac, M. J., Sargent, J. A., Powell, R. G., and Blackman, G. E. (1971).

Studies on foliar penetration. VIII. Effects of chlorination on the movement of phenoxyacetic and benzoic acids through cuticle isolated from the fruits of *Lycopersicon esculentum* L. *J. Exp. Bot.* **22**, 598.

Burrichter, E., Amelunxen, F., Vahl, J., and Giele, T. (1968). Pollen-und Sporen-untersuchungen mit dem Oberflächen—Rasterelektronenmikroskop. *Z. Pflanzenphysiol.* **59**, 226.

Campbell, R. (1972). Electron microscopy of the epidermis and cuticle of the needles of *Pinus nigra* var. *maritima* in relation to infection by *Lophodermella sulcigena. Ann. Bot.* **36**, 307.

Chapman, B. (1967). Polystyrene replicas for scanning reflexion electron microscopy. *Nature* **216**, 1347.

Conti, G. G., and Locci, R. (1972). Leaf surface alterations of *Nicotiana glutinosa* connected with mechanical inoculation of tobacco mosaic virus. *Riv. Patol. Veg.* (Padova), Ser. IV, **8**, 85.

Cross, P. M. (1972a). Specimen preparation. In: *The Use of the Scanning Electron Microscope* (Hearle, J. W. S., Sparrow, J. T., and Cross, P. M., eds.), p. 75. Pergamon Press, Oxford.

Cross, P. M. (1972b). Procedures for using a scanning electron microscope. In: *The Use of the Scanning Electron Microscope* (Hearle, J. W. S., Sparrow, J. T., and Cross, P. M., eds.), p. 87. Pergamon Press, Oxford.

Davis, D. G. (1971). Scanning electron microscope studies of wax formations on leaves of higher plants. *Can. J. Bot.* **49**, 543.

Echlin, P. (1968). The use of the scanning reflection electron microscope in the study of plant and microbial material. *J. Roy. Micros. Soc.* **88**, 407.

Echlin, P. (1971). Preparation of labile biological material for examination in the scanning electron microscope. In: *Scanning electron microscopy: Systematic and Evolutionary Applications.* The Systematics Association Special Volume 4 (Heywood, V. H., ed.), p. 307. Academic Press, London.

Echlin, P. (1972). Applications to biological materials. In: *The Use of the Scanning Electron Microscope* (Hearle, J. W. S., Sparrow, J. T., and Cross, P. M., eds.), p. 177. Pergamon Press, Oxford.

Echlin, P., and Hyde, P. J. W. (1972). The rationale and mode of application of thin films to non-conducting materials. In: *Proc. 5th Ann. Scan. Electron Micros. Symp.*, p. 137. I.I.T. Research Institute, Chicago.

Echlin, P., Paden, R., Dronzek, B., and Wayte, R. (1970). Scanning electron microscopy of labile biological materials maintained under controlled conditions. In: *Proc. 3rd Ann. Scan. Electron Micros. Symp.*, p. 50. I.I.T. Research Institute, Chicago.

Einert, A. H., De Hertogh, A. A., Rasmussen, H. P., and Shull, V. (1970). Scanning electron microscope studies of apices of *Lilium longiflorum* for determining floral initiation and differentiation. *J. Amer. Soc. Hort. Sci.* **95**, 5.

Falk, R. H., Gifford, E. M., and Cutter, E. G. (1970). Scanning electron microscopy of developing plant organs. *Science* **168**, 1471.

Falk, R. H., Gifford, E. M., and Cutter, E. G. (1971). The effect of various fixation schedules on the scanning microscope image of *Tropaeolum majus. Amer. J. Bot.* **58**, 676.

Fisher, R. W., and Corke, A. T. K. (1971). Infection of Yarlington Mill fruit by the apple scab fungus. *Can. J. Plant Sci.* **51**, 535.

Gibson, R. W. (1971). Glandular hairs providing resistance to aphids in certain wild potato species. *Ann. Appl. Biol.* **68**, 113.

Hanover, J. W., and Reicosky, D. A. (1971). Surface wax deposits on foliage of *Picea pungens* and other conifers. *Amer. J. Bot.* **58**, 681.

Hearle, J. W. S., Lomas, B., and Sparrow, J. T. (1970). The selection of conditions for examination of specimens in a scanning electron microscope. *J. Microscopy* **92**, 205.

Heslop-Harrison, Y. (1970). Scanning electron microscopy of fresh leaves of *Pinguicula*. *Science* **167**, 172.

Heslop-Harrison, Y., and Heslop-Harrison, J. (1969). Scanning electron microscopy of leaf surfaces. In: *Proc. 2nd Ann. Scan. Electron Micros. Symp.*, p. 117. I.I.T. Research Institute, Chicago.

Heywood, V. H. (1969). Scanning electron microscopy in the study of plant materials. *Micron* **1**, 1.

Hilliard, J. E. (1972). Quantitative analysis of scanning electron micrographs. *J. Microscopy* **95**, 45.

Holloway, P. J. (1967). Studies of the wettability of leaf surfaces. Ph.D. thesis, University of London.

Holloway, P. J. (1971). The chemical and physical characteristics of leaf surfaces. In: *Ecology of Leaf Surface Micro-organisms* (Preece, T. F., and Dickinson, C. H., eds.), p. 39. Academic Press, London.

Holloway, P. J., and Baker, E. A. (1968). Isolation of plant cuticles with zinc chloride-hydrochloric acid solution. *Plant Physiol.* **43**, 1878.

Huelin, F. E., and Gallop, R. A. (1951). Studies of the natural coating of apples. I. Preparation and properties of fractions. *Aust. J. Sci. Res.*, Ser. B., **4**, 526.

Hull, H. M. (1970). Leaf structure as related to absorption of pesticides and other compounds. In: *Residue Reviews* (Gunther, J. D., ed.), Vol. 31, p. 1. Springer-Verlag, Berlin.

Idle, D. B. (1969). Scanning electron microscopy of leaf surface replicas and the measurement of stomatal aperture. *Ann. Bot.* **33**, 75.

Jeffree, C. E., Johnson, R. P. C., and Jarvis, P. G. (1971). Epicuticular wax in the stomatal antechamber of Sitka spruce and its effect on the diffusion of water vapour and carbon dioxide. *Planta* **98**, 1.

Jensen, W. (1972). A study of the outer bark of birch (*Betula verrucosa*) and cork oak (*Quercus suber*) by scanning electron microscopy. *An. Quím. R. Soc. Esp. Fís. Quím.* **68**, 871.

Johnson, R. P. C., and Jeffree, C. E. (1970). Negative stain in wax tubes from the surface of Sitka spruce leaves. *Planta* **95**, 179.

Lane, G. S. (1972). Dimensional measurements. In: *The Use of the Scanning Electron Microscope* (Hearle, J. W. S., Sparrow, J. T., and Cross, P. M., eds.), p. 219. Pergamon Press, Oxford.

Lange, R. T. (1969). Concerning the morphology of isolated plant cuticles. *New Phytol.* **68**, 423.

Locci, R. (1969). Direct observation by scanning electron microscopy of the invasion of grapevine leaf tissues by *Plasmospara viticola. Riv. Patol. Veg.* (Padova), Ser. IV, **5**, 199.

Locci, R. (1971a). Preliminary observations by scanning electron microscopy on *Cycloconium oleaginum* Cast. *Riv. Patol. Veg.* (Padova), Ser. IV, **7**, 3.

Locci, R. (1971b). Osservazioni microscopiche relative all' infezione di tessuti fogliam di riso causata da *Helminthosporium oryzae. Il Riso* **20**, 225.

Locci, R., and Bisiach, M. (1971). Scanning electron microscopy of the invasion of leaf tissues by the apple scab fungus. *Riv. Patol. Veg.* (Padova), Ser. IV, **7**, 15.

Locci, R., and Quaroni, S. (1971). Scanning electron microscopy detected maize leaf modifications caused by *Helminthosporium maydis* and other micro-organisms. *Riv. Patol. Veg.* (Padova), Ser. IV, **7**, 109.

Martin, J. T., and Juniper, B. E. (1970). *The Cuticles of Plants.* Edward Arnold, London.

Muir, M. D., and Rampley, D. N. (1969). The effect of the electron beam on various mounting and coating media in scanning electron microscopy. *J. Microscopy* **90**, 145.

Murr, L. E. (1970). Scanning electron microscope studies of plant-leaf damage in an electrostatic field. In: *Proc. 3rd Ann. Scan. Electron Micros. Symp.*, p. 155. I.I.T. Research Institute, Chicago.

Newton, L. E. (1972). Taxonomic use of the cuticular surface features in the genus *Aloe* (Liliaceae). *J. Linn. Soc. (Bot)* **65**, 335.

Northrop, D. C. (1972). The interaction of electrons with solids. In: *The Use of the Scanning Electron Microscope* (Hearle, J. W. S., Sparrow, J. T., and Cross, P. M., eds.), p. 24. Pergamon Press, Oxford.

Orgell, W. H. (1955). The isolation of plant cuticles with pectic enzymes. *Plant Physiol.* **30**, 78.

Panessa, B. J., and Gennaro, J. F. (1972). Preparation of fragile botanical tissues and examination of intracellular contents by S.E.M. In: *Proc. 5th Ann. Scan. Electron Micros. Symp.*, p. 327. I.I.T. Research Institute, Chicago.

Pugh, G. J. F., and Buckley, N. G. (1971). The leaf surface as a substrate for colonization by fungi. In: *Ecology of Leaf Surface Micro-organisms* (Preece, T. F., and Dickinson, C. H., eds.), p. 431. Academic Press, London.

Purnell, T. J. (1971). Effects of pre-inoculation washing of leaves with water on subsequent infections by *Erysiphe cruciferarum.* In: *Ecology of Leaf Surface Micro-organisms* (Preece, T. F., and Dickinson, C. H., eds.), p. 269. Academic Press, London.

Rentschler, I. (1971). Die Wasserbenetzbarkeit von Blattoberflächen und ihre submikroskopische Wachsstruktur. *Planta* **96**, 119.

Reyre, Y. (1968). La sculpture de l'exine des pollen des Gymnospermes et des Chlamydospermes et son utilization dans l'identification des pollen fossiles. *Pollen et Spores* **10**, 197.

Royle, D. J., and Thomas, G. G. (1971). Observations with the scanning electron microscope on the early stages of hop leaf infection by *Pseudoperonospora humuli. Physiol. Plant Pathol.* **1**, 345.

Sacalis, J. N. (1972). Rose petal surfaces. *Hortscience* **7,** 2.

Schwerdtfeger, G. (1972). Die Oberfläche der Apfelfrucht in Aufnahmen mit dem Raster-Elektronen-Mikroskop und Elektronen-Mikroskop. *Der Erwerbsobstbau* **14,** 17.

Still, G. G., Davis, D. G., and Zander, G. L. (1970). Plant epicuticular lipids: alteration by herbicidal carbamates. *Plant Physiol.* **46,** 307.

Troughton, J., and Donaldson, L. A. (1972). *Probing Plant Structure.* Chapman and Hall, London.

Verdus, M.-C. (1969). L'epiderme cotyledonaire d'*Euphorbia corsica* Req. *C. r. hebd. Séances Acad. Sci., Paris,* Ser. D., **268,** 793.

Verdus, M.-C. (1970). Contribution a l'etude des plantules d'Euphorbiacees. *Travaux du Laboratoire Forestier de Toulouse* **8,** article 9, 1.

Welford, W. T. (1972). On the relationship between the modes of image formation in scanning microscopy and conventional microscopy. *J. Microscopy* **96,** 105.

Zachariah, K., and Pasternak, J. (1970). Processing soft tissues for scanning electron microscopy. Simplification of the freeze drying procedure. *Stain Technol.* **45,** 43.

9. PLANT CELL WALLS AND INTRACELLULAR STRUCTURES

Lewis G. Briarty

Botany Department, University of Nottingham, Nottingham, England

INTRODUCTION

The scanning electron microscope (SEM) was conceived for use in a reflection mode and until recently has been for the most part so used by biologists. This and the significant difficulties involved in specimen preparation have resulted in the production of vast numbers of aesthetically pleasing images of external surface features of organisms. However, very little work has been carried out on the intracellular features, where the magnification and depth of field of the instrument can be used to resolve problems involving three-dimensional cellular architecture. The main obstacle to such intracellular investigation lies in the difficulty of making accessible and isolating the particular structure being studied. Intracellular structures are by definition hidden, and so methods must be developed to reveal them.

The problems encountered differ widely depending upon the nature of the specimen, but in general they are of two types: those which involve the cutting or breaking of cell walls to reveal the interior of the cell and those which involve the isolation and delineation, within the fractured cell, of the component under investigation.

METHODS FOR OPENING CELLS

The simplest procedure to view the interior of cells in a tissue is simply to cut blocks or sections of the specimen and to look into those cells which have been cut open. In the case of firm tissues, this can be accomplished using a new razor blade; in the case of plant tissues, the cut cell walls are left fairly smooth. Jaques et al. (1965) used this technique to investigate three-dimensional spatial relationships in tissues, observing 1 mm³ blocks of lung and intestine which had been soaked in, or coated with a saturated solution of gold chloride (see also Siew, 1970).

One of the earliest observations of sectioned material with the SEM was carried out by McDonald et al. (1967a), who used paraffin-embedded and stained animal tissues to obtain information on the rigidity and bulk of tissue components. Further general comparisons of light microscopy and scanning electron microscopy applied to deparaffinized sections (McDonald et al., 1967b) showed that a fairly high resolution could be attained. Other tissues which have been examined by simple sectioning are barley grain (Pomeranz, 1972a and b), oat kernel (Pomeranz, 1972c), buckwheat kernel (Pomeranz, 1972d), triticale grain (Pomeranz, 1971), and wheat endosperm (Dronzek et al., 1972). In the fixed and dried sections of hen oviduct, Makita and Sandborn (1971) were able to identify the nucleus and intracellular secretion granules.

A more sophisticated means of revealing intracellular detail is provided by freeze-fracturing (see Nei, in this volume). This method involves freezing the tissue in liquid nitrogen (or, better, in Freon or a similar halogen-substituted hydrocarbon), fracturing it mechanically, freeze-drying it, and then vacuum-coating the fractured surface. Such a procedure has been used by Germinario and McAlear (1971) to study the structure of mouse retina.

When the specimen is too small or delicate to be cut mechanically, freeze-fracturing can be used as exemplified by work on Euglena (Guttman, 1971; Guttman and Styskal, 1971). In freeze-fractured lyophilized cells of Euglena, the authors claim to have resolved subcellular organelles including the endoplasmic reticulum. These authors also use a "rough" fracturing method, which involves sonicating the cells with glass beads.

An ingenious method for cryofracturing small fragments of tissue has been devised by Lim (1971). The tissue (in this case, parts of the inner ear) is suspended in a drop of 70% ethanol and dropped into liquid nitrogen. The drop fractures as a result of internal stresses, and reveals the inner cell surfaces.

METHODS FOR REVEALING PARTICULAR INTRACELLULAR STRUCTURES

Much of the early use of the SEM by botanists for looking into cells dealt with investigating the structure of wood (Resch and Blaschke, 1968; Richter *et al.*, 1968; Ishida and Ohtani, 1969; see also Borgin, in this volume). The situation is simple in studying wood, in that the cells are dead and without cytoplasmic contents and, in sectioned tissue, the inner walls of the xylem are open to inspection. However, when there is a need to observe the contents or the interior walls of living cells (i.e., cells containing protoplasts), significant problems arise in specimen preparation.

Two factors are important: possible distortion of the *in vivo* situation during tissue preparation, and the effectiveness of the methods used to expose the particular subcellular structure under investigation. The former is a complex problem, and has been discussed in other chapters in this volume. However, it must be stated that the time at which probably the most structural distortion can occur is during the removal of water. With the continuing requirement for finer structural preservation, the dehydration process has been refined from simple air-drying or dehydration in an alcohol series (Jaques *et al.*, 1965) through freeze-drying to critical point drying, and culminating in the examination of hydrated living material. For a more extensive treatment of this subject, the reader is referred to Echlin (1968 and 1972) and Boyde and Wood (1969).

The more intractable problem, however, is that of exposing selectively a particular intracellular structure for examination. When the aim is to examine the surface of the cytoplasm in vacuolated cells, it is possible to substitute polyethylene glycol for the water in the cell walls and cytoplasm (Idle, 1971). A similar approach was taken by Chambers and Hamilton (1973) to study wheat node transfer cells. They developed a method for "partial embedding" of the tissue in an epoxy resin, so that the water in the cells is replaced by polymerized resin. A particular advantage of this method is that the tissue, after examination in the SEM, can be reinfiltrated with resin, polymerized, and then sectioned for further study.

When a particular intracellular component is to be observed, however, additional problems arise which must be dealt with individually. The factor common to all these problems entails the removal of some parts of the cell while other parts are left intact. The possible means of doing this include mechanical dissection, ion beam etching, and chemical or enzymatic dissolution of the unwanted material.

Mechanical Dissection

This procedure has as yet not been applied to intracellular components. However, in one instance it has been used to demonstrate relationships at neuronal surfaces (Lewis, 1971). In this study, mechanical dissection was used both with and without prior treatment of the tissue with 5 mM EDTA (see also Zeevi and Lewis, 1970).

Ion Beam Etching

This procedure provides a controlled means for removing materials from the specimen; but it is nonspecific, and the results are not easy to interpret (Echlin, 1972; Carteaud, 1970).

Enzyme and Chemical Etching

These procedures have provided the greatest number of observations of cellular components, though even now only relatively solid structures can be successfully demonstrated. The methods for preparing various intra- and extracellular wall surfaces and some cytoplasmic components are presented below.

Intracellular Wall Surfaces. The studies of transfer cell structure and function (Pate *et al.*, 1969) led to the development of an enzyme etching method for preparing the peculiar wall labyrinth, which these cells possess, for study with the SEM. This is a three-dimensional structure at the limit of resolution of the light microscope, and is too complex to be understood easily from thin sections used for TEM work.

The method which follows has been briefly described elsewhere (Briarty, 1971), and has proved suitable for cleaning internal cell wall surfaces of a number of tissues. It is presented below as a basis for further experimentation.

(1) Cut the tissue into thin slices or into cubes (~1 mm on a side), using a new, degreased razor blade.

(2) Place the tissue specimens in enzyme solution in stoppered specimen vials and incubate at 37°C for 72 hr. The enzyme used is a filtered 1% w/v solution of bromelein (Sigma Chemical Co.) (an active protease derived from pineapple) in 0.1 M phosphate/citrate buffer (pH 4.5).

Ideally the preparation should be filter-sterilized; however, microbial contamination which occurs during the incubation can be washed off the specimen at a later stage.

(3) Remove the enzyme mixture from the specimen vials and replace with freshly prepared distilled water. Wash the specimens by agitating for 1 min with a Pasteur pipette, gently sucking the tissue blocks into the pipette and then blowing them out. The shearing forces involved in this washing help to remove the cell contents and contamination from the surface of the tissue. Carry out this washing procedure at least six times.

(4) Freeze the tissue blocks by plunging them, on pieces of aluminum foil, into liquid nitrogen; then transfer the foil and tissue to the cooled table of a histological freeze-dryer set to the lowest temperature. Freeze-dry the material.

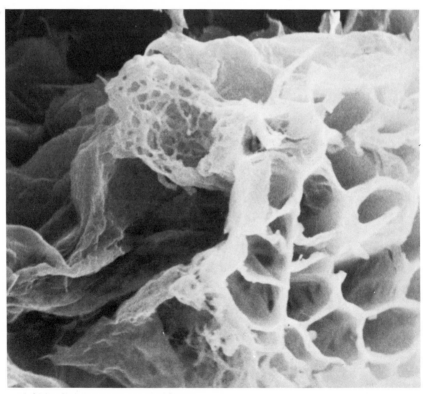

Fig. 9–1. A transfer cell (center) adjacent to a vascular bundle (right) from a root nodule of *Trifolium repens*, prepared using bromelein only. 4,495 X.

(5) Transfer the dry tissue blocks to SEM specimen stubs and attach them in the usual way with glue or conducting paint, and coat with gold. Root nodule transfer cells prepared by this method are shown in Fig 9–1.

The duration of digestion is rather critical; although a period of 24 hr does not allow sufficient digestion, longer durations result in microbial contamination, which is difficult to remove from the surface of the tissue.

The result of adding lipase (from hog pancreas, Sigma Chemical Co.) at 1% w/v was, in some cases, to improve the extent of the etching effect of bromelein at pH 7.0 (Fig. 9–2); usually the latter was not very effective when used alone at a neutral pH. The further addition of 1% w/v of Onozuka Macerozyme to the protease-lipase mixture at pH 4.5 appeared to cause some dissolution of the cell walls; and, though the cells remained still attached together, holes appeared in thin areas of walls (Fig. 9–3). At pH 7.0, this mixture of enzymes was ineffective in cleaning the cell walls (Fig. 9–4).

Other tissues which have been prepared using this method of wall cleaning include a number of legume seed cotyledons. The cells are packed with reserve products (starch and globulin protein in the cases studied), and cleaning is not complete. In some cases where only the tip of a cell is cut open, the contents remain fairly undigested. The starch is apparently not attacked, and comparatively full cells are found next to clean, empty ones (Fig. 9–5).

The cleaned cell walls can be examined for details of wall architecture; for example, in the cotyledon parenchyma cells of the two species of legumes (Figs. 9–6 and 9–7), the arrangement of pit fields and wall thickenings shows distinct differences which must be related to the flow of water and solutes through the cotyledons. Moreover, it should be possible to analyze these data quantitatively using the methods discussed by Hilliard (1972). It is fairly easy, using such cleared cell walls, to understand the three-dimensional structure of a tissue and the way in which cell shape affects the distribution of intracellular spaces (Fig. 9–8).

Extracellular Wall Surfaces. As with the observation of cell interiors, the problem encountered in studying the outer surfaces of cells is one of accessibility. In this case, enzyme treatment can also be useful. The following method was suggested by M. Stein for revealing details of cell wall surfaces which are normally obscured at the cell/cell interface.

(1) Seeds of *Phaseolus vulgaris* are soaked at 40°C until fully imbibed (3 to 4 hr).

(2) Small blocks of cotyledon tissue are placed in a 3% aqueous solution of Rohament P (a macerase) for 5 to 24 hr at room temperature.

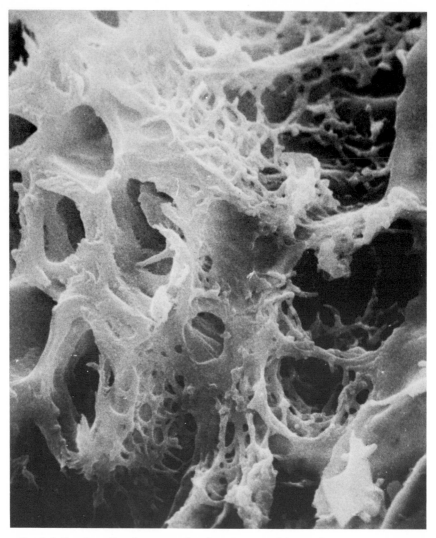

Fig. 9–2. Transfer cells adjacent to xylem from a root nodule vascular bundle in *T. repens,* prepared using lipase and bromelein. 4,266 X.

(3) The tissue blocks are rinsed, freeze-dried, then mounted on specimen stubs, and coated according to the standard procedures.

Part of the outer surface of a cotyledon cell is shown in Fig. 9–9; the pit fields protrude from the general wall material. In Fig. 9–10, the

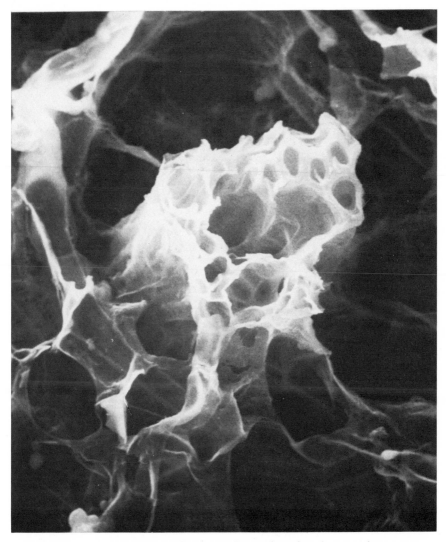

Fig. 9–3. Root nodule vascular bundle, prepared using bromelein, lipase, and macerozyme; the unthickened cells around the vascular bundle are distorted, with breaks in their walls. 2,296 X.

structure of the outer side of the pit field is clear, and the protuberances on the surface presumably represent the sites of plasmodesmata.

A number of other degradative methods have been devised for examining outer surfaces of cell walls. Nozawa and Ito (1970) investigated the structure of the spore wall of *Microsporum cookei* using snail gut enzyme, the best conditions being a solution of 10 mg of enzyme per ml of 0.05 M pH 5.8 citrate/phosphate buffer, with 0.01% cystein hydrochloride added. Treatments of up to 24 hr removed protuberances from the walls and decreased their rigidity. In studies on pollen wall formation, Heslop-Harrison (1968) used the callase activity of Onozuka Cellulase P500 as well as acetolysis and cuprammonium hydroxide treatment to investigate the various wall layers.

The effect of a number of maceration techniques (acetic acid with hydrogen peroxide, nitric acid, triethanolamine) on fiber structure of woods of varying hardness was studied by van Huynh-Long and Homès

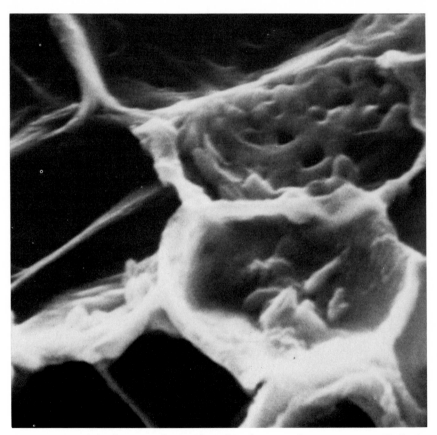

Fig. 9–4. Transfer cell, prepared using bromelein, lipase, and macerozyme at pH 7; the cytoplasm does not appear to have been removed from the wall protuberances. 11,484 X.

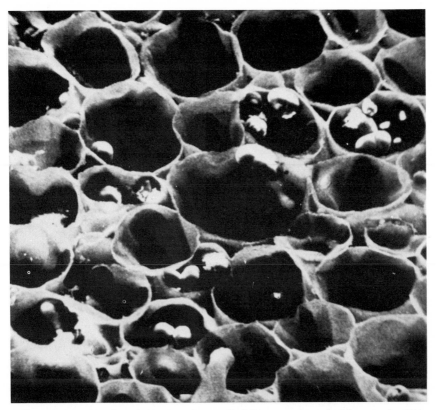

Fig. 9–5. A section of cotyledon tissue from seed of *Pisum arvense* soaked in water for 24 hr and treated with bromelein for 72 hr. 328 X.

(1971). They concluded that triethanolamine treatment was most effective in dissociating the fibers. Another study of wood structure (Scurfield and Silva, 1969) examined the wall architecture of conifer reaction wood tracheids before and after treatment with a variety of acid and alkaline delignifying agents.

In order to examine the structure of cuticle, Baker (1970) removed the wax from isolated plant cuticles using chloroform and then observed the cellulose after treatment with $ZnCl_2/HCl$, oxalic acid, and ammonium oxalate. Finally, an aspect of wall structure which is revealed without any direct treatment is the anatomy of the leaf abscission zone. This was described by Rogers (1971), who noted that in Valencia orange fruit the abscission zone showed less cell deformation after ethylene treatment of the ripe fruit.

Cytoplasmic Details. Perhaps the only demonstration so far of specific etching by enzymes to reveal cytoplasmic detail has been carried out on wheat grains. The following method was suggested by A. D. Evers. The structure of starch grains in sections of the endosperm and pericarp cells

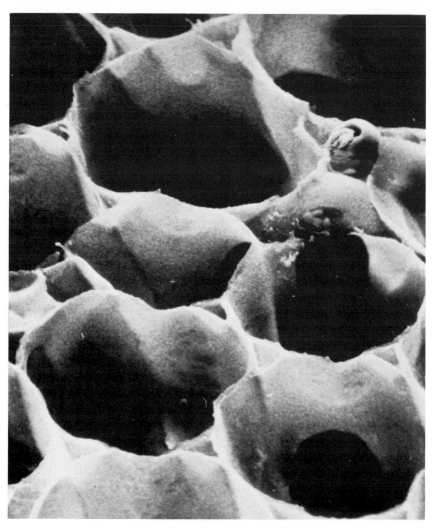

Fig. 9–6. A cotyledon section from *P. arvense,* treated as described for Fig. 9–5. 845 X.

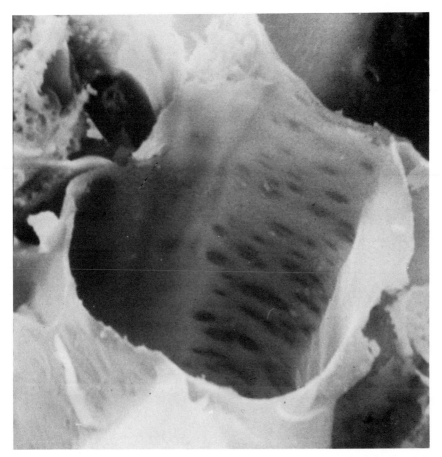

Fig. 9–7. A cotyledon section from seed of *Vicia faba,* treated as described for Fig. 9–5. 1,800 X.

is obscured by the investing cytoplasm. The problem is to remove the cytoplasm without disturbing the starch, and this is achieved by the following enzyme etching method.

(1) 20 μm thick cryostat sections of wheat grain are picked up on small membrane filters.

(2) The filters are placed, section uppermost, on a pad of 6 filter papers soaked in a 1% w/v pronase solution containing 0.1% Thiomersal (BDH) to inhibit microbial activity and incubated in a Petri dish at 40°C for 48 hr. The sections are then air-dried, attached to SEM stubs, and coated according to the standard procedures. Figs. 9–11 and 9–12

show pericarp cells before and after pronase treatment respectively, and Fig. 9–13 shows developing endosperm material. The starch grains are clearly delineated, distinct from the cytoplasm. Further examples of enzyme etching which occur *in vivo* are provided by studies of the effect of α-amylase on starch during germination in cereal endosperms (Pomeranz, 1972b; Dronzek *et al.*, 1972). The starch grains are attacked in particular areas by the enzyme, and the grain sometimes splits to reveal its inner lamellar structure (Fig. 9–14) (Evers and McDermott, 1970).

The methods which have been developed for studying cell walls and intracellular structure with the SEM reflect the diversity of the problems involved. The situations mentioned above present solutions to some

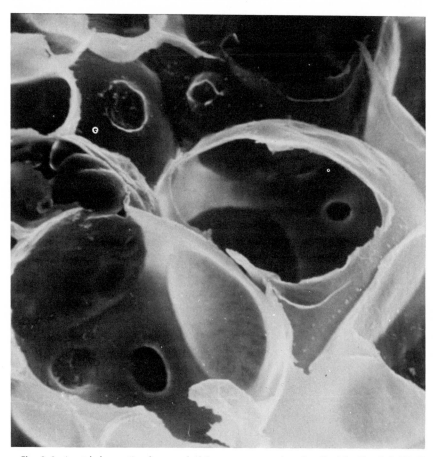

Fig. 9–8. A cotyledon section from seed of *P. arvense*, treated as described for Fig. 9–5. 771 X.

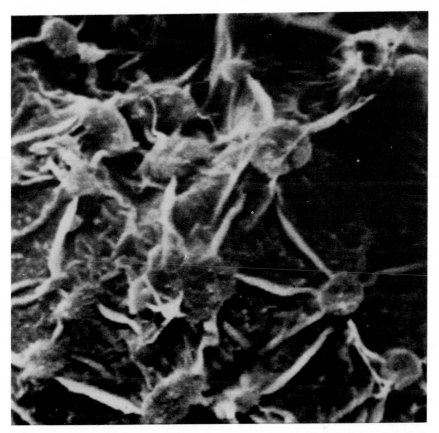

Fig. 9–9. Surface view of the exterior wall of a cotyledon parenchyma cell from *P. vulgaris* showing pit fields. No macerase treatment. Courtesy M. Stein. 3,750 X.

of the difficulties arising from the use of this relatively new instrument to help elucidate the relationships between structure and function in plant cells.

I am grateful to A. D. Evers (Flour Milling and Baking Research Association, Chorleywood, Hertfordshire) and Dr. M. Stein (School of Agriculture, University of Nottingham) who made their work available to me before publication.

REFERENCES

Baker, E. A. (1970). The morphology and composition of isolated plant cuticles. *New Phytol.* **69,** 1053.

Boyde, A., and Wood, C. (1969). Preparations of animal tissues for surface-scanning electron microscopy. *J. Microscopy* **90,** 221.

Briarty, L. G. (1971). A method for preparing living plant cell walls for light and electron microscopy. *J. Microscopy* **94,** 181.

Carteaud, A. J. P. (1970). Applications de techniques physiques à l'étude d' échantillons biologiques en microscopie électronique par balayage. *Microscopie Electronique* **1,** 473. *Proc. 7ème. Cong. Internat. Micros. Electronique.* Grenoble.

Chambers, T. C., and Hamilton, C. D. (1973). Scanning electron microscopy of transfer cells: A new method for preparation of plant tissues. *J. Microscopy* **99,** 65.

Dronzek, B. L., Hwang, P., and Bushuk, W. (1972). Scanning electron microscopy of starch from sprouted wheat. *Cereal Chem.* **49,** 232.

Echlin, P. (1968). The use of the scanning reflection electron microscope in the study of plant and microbial material. *J. Roy Micros. Soc.* **88,** 407.

Echlin, P. (1972). Applications to biological materials. In: *The Use of the Scanning Electron Microscope* (Hearle, J. W. S., Sparrow, J. T., and Cross, P. M., eds.), pp. 177–202. Pergamon Press, Oxford.

Evers, A. D., and McDermott, E. E. (1970). Scanning electron microscopy of

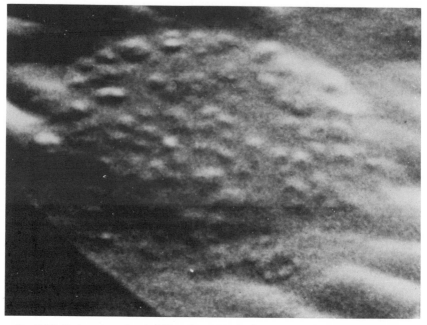

Fig. 9–10. Surface view of a pit field at the exterior wall of a parenchyma cell from *P. vulgaris,* treated with macerase. Courtesy M. Stein. 3,700 X.

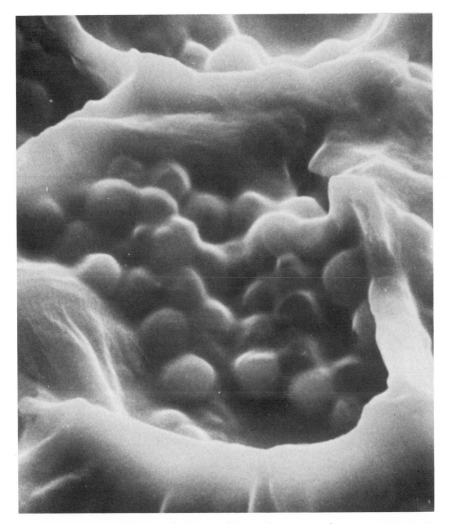

Fig. 9–11. A section of young developing wheat seed pericarp without pronase treatment. Courtesy A. D. Evers and the Cambridge Instrument Co. 7,150 X.

wheat starch. II. Structure of granules modified by alpha-amylosis-preliminary report. *Die Stärke* **22**, 23.

Germinario, L. T., and McAlear, J. H. (1971). Preparation of tissue for scanning electron microscopy: Freeze-fracturing as a technique for enhancing visibility of structural relationships. *Stain Technol.* **46**, 249.

Guttman, H. N. (1971). Internal cellular details of *Euglena gracilis* visualized by scanning electron microscopy. *Science* **171**, 290.

Guttman, H. N., and Styskal, R. C. (1971). Preparation of suspended cells for scanning electron microscope examination of internal cellular structures. *Proc. 4th Ann. Scanning Electron Microscopy Symp.*, p. 265–71. I.T.T. Research Institute, Chicago.

Heslop-Harrison, J. (1968). Wall development within the microspore tetrad of *Lilium longiflorum. Can. J. Bot.* **46,** 1185.

Hilliard, J. E. (1972). Quantitative analysis of scanning electron micrographs. *J. Microscopy* **95,** 45.

van Huynh-Long, and Homès, J. (1971). Quelques considérations sur les aspects morphologiques des fibres de bois au cours de la macération. *Bull. Soc. r. Bot. Belg.* **104,** 151.

Idle, D. B. (1971). Preparation of plant material for scanning electron microscopy. *J. Microscopy* **93,** 77.

Ishida, S., and Ohtani, J. (1969). Observations of bordered pits in softwood tracheid made with the scanning electron microscope. *JEOL News* **7B,** 3.

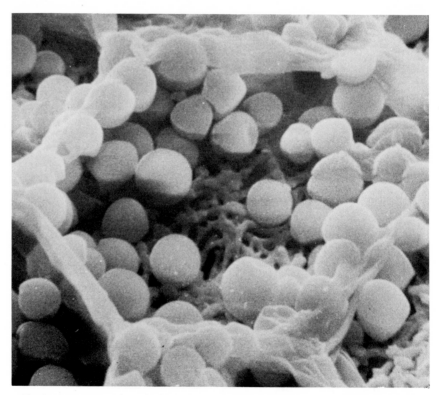

Fig. 9–12. A section of young developing wheat seed pericarp after pronase treatment. Courtesy A. D. Evers and the Cambridge Instrument Co. 8,200 X.

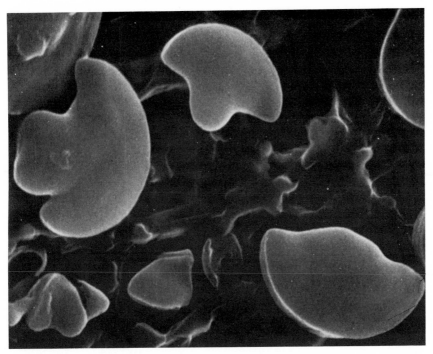

Fig. 9–13. Starch grains in developing endosperm cells of wheat seed after pronase treatment. Courtesy A. D. Evers and the Cambridge Instrument Co. 3,373 X.

Jaques, W. E., Coalson, J., and Zervins, A. (1965). Application of the scanning electron microscope to human tissues. *Exp. Mol. Pathol.* **4,** 576.

Lewis, E. R. (1971). Studying neuronal architecture and organization with the scanning electron microscope. *Proc. 4th Ann. Scanning Electron Microscopy Symp.*, pp. 281–88, I.T.T. Research Institute, Chicago.

Lim, D. J. (1971). Scanning electron microscopic observation on non-mechanically cryofractured biological tissue. *Proc. 4th Ann. Scanning Electron Microscopy Symp.*, pp. 359–64, I.T.T. Research Institute, Chicago.

McDonald, L. W., Pease, R. F. W., and Hayes, T. L. (1967a). The use of the scanning electron microscope for the examination of sectioned tissues. *Electron Microscopy*, pp. 254–55. *Proc. E.M.S.A. 25th Anniversary Meeting.* Claitor's Publishing Division, Baton Rouge, La.

McDonald, L. W., Pease, R. F. W., and Hayes, T. L. (1967b). Scanning electron microscopy of sectioned tissues. *Lab. Invest.* **16,** 532.

Makita, T., and Sandborn, E. B. (1971). Identification of intracellular components by scanning electron microscopy. *Expl. Cell Res.* **67,** 211.

Nozawa, Y., and Ito, Y. (1970). Digestion process of *Microsporum cookei* spore wall by snail enzyme: Scanning electron microscope investigation. *Experientia* **26**, 801.

Pate, J. S., Gunning, B. E. S., and Briarty, L. G. (1969). Ultrastructure and functioning of the transport system of the leguminous root nodule. *Planta* **85**, 11.

Pomeranz, Y. (1971). Functional characteristics of triticale, a man-made cereal. *Wallerstein Laboratories Communications* **34**, 175.

Pomeranz, Y. (1972a). Determination of the structure of barley kernel by scanning electron microscopy. *Cereal Chem.* **49**, 1.

Pomeranz, Y. (1972b). Scanning electron microscopy of the endosperm of malted barley. *Cereal Chem.* **49**, 5, 18.

Pomeranz, Y. (1972c). Scanning electron microscopy of the oat kernel. *Cereal Chem.* **49**, 20.

Pomeranz, Y. (1972d). Scanning electron microscopy of the buckwheat kernel. *Cereal Chem.* **49**, 23.

Resch, A., and Blaschke, R. (1968). Über die Anwendung des Raster-Elektronenmikroskopes in der Holzanatomie. *Planta* **78**, 85.

Richter, I.-E., Vogel, K., and Huber, H.-J. (1968). Die Untersuchung unfixier-

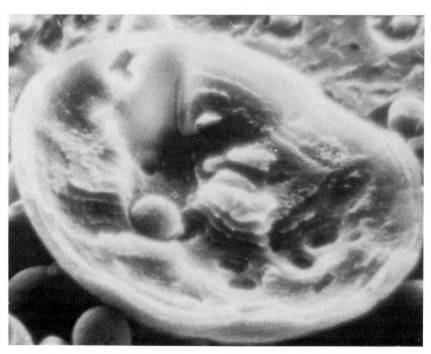

Fig. 9–14. A starch grain from endosperm of a germinating wheat seed showing the results of attack by α-amylase. The grain has split along the equatorial plane of weakness. Specimen mounted on a membrane filter. Courtesy A. D. Evers and the Cambridge Instrument Co. 40,523 X.

ter pflanzlicher Objekte mit dem Raster-Elektronenmikroskope "Stereoscan." *Z. wiss. Mikrosk.* **69**, 94.

Rogers, B. J. (1971). Scanning electron microscopy of abscission-zone surfaces of Valencia-orange fruit. *Planta* **97**, 358.

Scurfield, G., and Silva, S. (1969). The structure of reaction wood as indicated by scanning electron microscopy. *Aust. J. Bot.* **17**, 391.

Siew, S. (1970). The application of the scanning electron microscope in the study of cardiac and pulmonary tissue. *Microscopie Electronique,* **1**, 479. *Proc. 7ème. Cong. Internat. Micros. Electronique.* Grenoble.

Zeevi, Y. Y., and Lewis, E. R. (1970). A new technique for exposing neuronal surfaces for viewing with the scanning electron microscope. *Microscopie Electronique* **1**, 481. *Proc. 7ème. Cong. Internat. Micros. Electronique.* Grenoble.

10. INTRACELLULAR STRUCTURES

Barbara J. Panessa

St. Vincent's Hospital, School of Nursing, New York

and

Joseph F. Gennaro, Jr.

Laboratory of Cellular Biology, New York University, New York

INTRODUCTION

It is anticipated that higher-resolution scanning electron microscope (SEM) images will be attained in the near future. Already reports indicate the achievement of 25 to 50 Å resolution by using modified instruments. It seems only a matter of time before SEM resolution will approach that of the transmission electron microscope (TEM). With this prospect in view and the information on surfaces already provided by SEM, the next logical step is to examine the disposition of intracellular organelles. In order to accomplish this, it is essential that the morphology of the cell interior be accessible to the electron beam.

The nature of electron micrographs produced in the emissive mode, which has provided us with elegant data concerning cell surface details, is such that it is not possible to visualize structures which are located deeper than 50 Å from the surface (Beaman and Isasi, 1972). This type of exposure may be achieved by cleaving open fixed or frozen cells (Guttman and Styskal, 1971; Wodzicki and Humphreys, 1973), but it is also possible to use plant cells in which the central or tonoplast vacuole is so large that organelles are distributed on the cell walls and project into the vacuolar space. In such a system, the organelles are delineated by the limiting membrane of the tonoplast vacuole (Fig. 10–1). As is the case with transmission electron microscopy, specimen preparation pre-

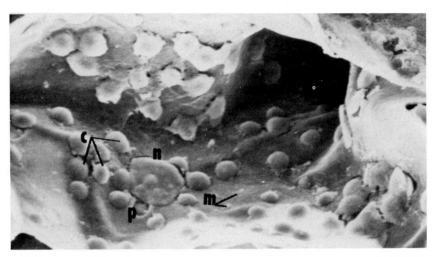

Fig. 10–1. Gold-palladium-coated interior of a mesophyll cell of *Sarracenia purpurea*. Note the smooth appearance of the tonoplast membrane deformed by cytoplasmic organelles which project into the vacuole. *n*, nucleus; *m*, mitochondria; *c*, chloroplasts; *p*, primary pit field; *cw*, cell wall, 10 kV. 1,900×.

sents a far more serious obstacle to obtain necessary information than does instrument capability.

COATED SPECIMENS

Plant specimens are well suited for intracellular observation because of the large tonoplast vacuoles found in some mesophyll and parenchymous cells (Fig. 10–2). By fixing small pieces of leaf tissue (2 to 4 mm²) for days to weeks in a slightly hypertonic aldehyde fixative, such as glutaraldehyde (3 to 6% in 0.1 M cacodylate buffer, pH 7.4), adequate penetration of the fixative can be achieved. An advantage of the long duration of fixation is the eventual hardening of the specimen. In this state, it is less likely to be damaged during subsequent processing, better withstands the effects of the electron beam and vacuum in the specimen chamber, and can be retrimmed with a razor or scalpel blade to expose desired structures without crushing or distortion.

The fixed tissue is rinsed in 0.1 M cacodylate buffer to remove residual fixative. It is desirable to trim the edges of the tissue block and remove portions which have been damaged during excision and fixation. Postfixation is carried out with 2% aqueous uranyl acetate for 24 hr at room temperature. This heavy metal has been shown to bind tissue lipids

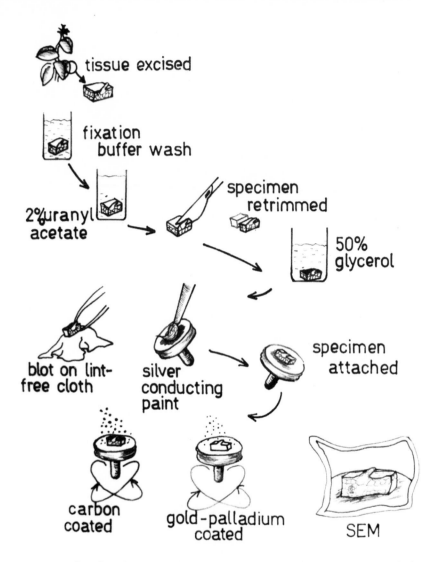

Fig. 10–2. Flow chart diagram of tissue preparation for SEM examination of glycerol-substituted, metal-coated specimens. See Figs. 10–1 and 10–3.

(Mumaw and Munger, 1971), and penetrates into the tissue more rapidly than does conventionally employed osmium tetroxide. In addition, uranyl acetate does not blacken the tissue, a characteristic of osmication that makes orientation of the specimen difficult. Since the tissues examined with the scanning electron microscope may be subsequently embedded and sectioned for examination with the TEM, the uranyl acetate treatment provides for contrast enhancement. There is some indication that the introduction of high atomic number materials into the specimen increases tissue conductivity under the electron beam, producing a higher contrast and consequently better imaging (Pfefferkorn, 1970; Idle, 1971).

After postfixation with uranyl acetate, the specimen is washed in the buffer for a few minutes, followed by two washes with distilled water (½ to 1 min each). The specimen is then placed in aqueous 50% glycerol for 30 min (this duration is sufficient for the glycerol to infiltrate the specimen) or can be stored for several weeks in 50% glycerol in the cold. A reliable criterion for proper glycerol infiltration is the loss of buoyancy of the tissue specimen.

Immediately before examination with the SEM, the tissue specimen is removed from glycerol and placed on a lint-free cloth. This permits drainage of excess glycerol from the surfaces, and removes the glycerol from the bottom which is attached to the stub or specimen support. Silver conducting paint is applied liberally to the stub as a bubble-free film and allowed to dry to stickiness in order to prevent the film from creeping up the sides of the tissue specimen by capillarity. The specimen is oriented on the painted stub and left to dry for a few minutes in a clean area (Panessa and Gennaro, 1972).

After the paint has been dried, the specimen support may be introduced into the chamber of the vacuum evaporator. The use of a rotary tilt stage during vacuum evaporation of the conducting material onto the specimen surface promotes an even and complete metal film. A thin layer of carbon ($\sim$50 Å) is slowly evaporated onto the specimen at 10^{-4} torr. The carbon tends to adhere to irregular tissue surfaces, and provides a more uniform, stable substrate on which a metal film may be deposited (Echlin and Hyde, 1972). Without a carbon stabilizing layer, holes and cracks frequently occur in the metal during specimen examination. A metal film of 100 to 250 Å thickness should be evaporated onto the carbon film at 10^{-5} torr. It has been our experience and that of others (Echlin and Hyde, 1972; Boyde, 1972) that gold-palladium is preferable to other coating materials (Ag, Al, Au, C, Cr, Cu, Pd, and Pt). Its effectiveness is due to its ability to produce satisfactory secondary emission under electron beams of very different intensities, and it can be used to form strong, thin films which are resistant to oxidation. In addition, gold-palladium has a smaller grain size on deposition than all of the coating metals mentioned except platinum. Care must be taken, however, in positioning the gold-palladium electrode far enough from the tissue during evaporation to prevent heat radiation damage of the specimen. Also, a metal "heat shield" can be placed between the evaporating electrode and the specimen.

Specimens prepared using the above-mentioned technique withstand the effects of vacuum and beam well, provided they have been sufficiently hardened during the fixation process (Fig. 10–3). Glycerol infiltration also adds to the stability of the specimen. The water in the tissue specimen is partially replaced and possibly bound by the infiltrated glycerol. The low vapor pressure of the glycerol prevents its evaporation during the deposition of the carbon and metal films which "seal" tissue surface. Well-prepared specimens cut open from several days to months after examination with the SEM were moist and showed no signs of decay or collapse. Glycerol has been used in conjunction with evaporation techniques in the manufacture of carbon-coated Formvar films without contamination difficulties in one of the oldest techniques of electron microscopy. Although it has been a matter of controversy, the authors have found no indication that partial replacement of water in the tissue by glycerol substitution adds to microscope contamination.

If contamination is a consideration, specimen infiltration with poly-

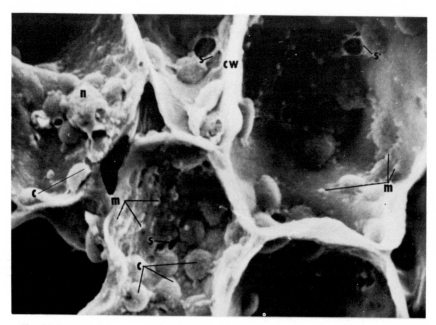

Fig. 10–3. Intracellular organelles are easily visible in the interior of epidermal cells in the leaf of *Sarracenia*. Some cytoplasmic strands (*s*) can be seen. The cytoplasm has a smooth appearance characteristic of the metal filled specimens. 10 kV, 2,000×

ethylene glycol (Idle, 1971) which is water-miscible and has a vapor pressure orders of magnitude lower than that of glycerol, may be preferred. Padawar (1971) has described a preparation procedure for mast cell observation by the SEM in which tissue water is replaced by silicone substitution. However, both silicone and polyethylene glycol substitution include acetone dehydration (Idle suggests the use of acetone/PEG-400 mixtures), which has been reported to produce tissue shrinkage and distortion (Boyde, 1972).

VERIFICATION OF SEM IMAGES

It is essential to verify that the structures observed with the SEM are real. It is always conceivable, especially at high magnifications, that the preparatory procedures including coating have produced artifacts which lead to false representation. We have found that verification requires removal of the scanning specimen from the stub, followed by its embedding, sectioning, and examination with the light and/or transmission electron microscope (Panessa and Gennaro, 1972). Specimens which have been scanned may be removed from the stub with a razor blade (Fig. 10–4). Care should be taken to remove all traces of silver conducting paint from the bottom of the tissue, since its incomplete removal results in the infiltration of tissue with electron-dense metal particles during the embedding procedure.

Specimens are passed through 3 to 4 washes of absolute acetone in order to remove any water that may have remained in the tissue as well as glycerol. The specimens are then gradually infiltrated with epoxy resin by adding the catalyzed resin dropwise to the acetone-immersed tissue. This is carried out gradually with constant swirling and should take 10 to 20 min to triple the volume of the mixture. Once the concentration of the resin has reached this stage (~75–80% epoxy resin), the mixture can be removed and replaced with fresh catalyzed resin. To insure adequate penetration of the mixture throughout the tissue and remove residual acetone, the 100% resin with the tissue is left at room temperature overnight. To avoid introducing dirt into the resin during this procedure but allow the volatilization of residual acetone, the vial is covered with a clean piece of lint-free filter paper.

In order to insure infiltration of the resin and its complete polymerization, the embedding medium is made with half the quantity of accelerator usually employed. This mixture containing the tissue is transferred to a shaker at room temperature for 2 days. Each day the resin is changed. Again, loss of tissue buoyancy usually indicates satisfactory infiltration. After thorough infiltration, the resin may be polymerized. Due to the

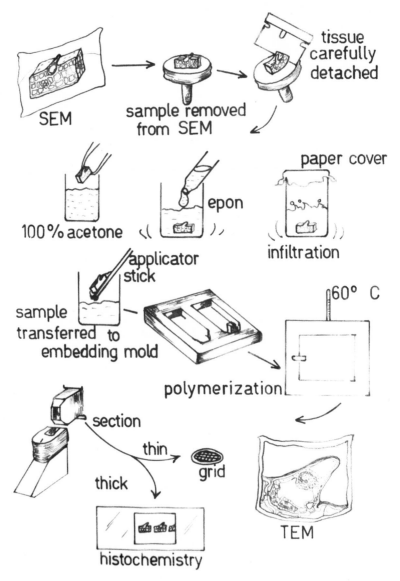

Fig. 10–4. Flow chart of the SEM specimen reembeddment technique employed in the verification of SEM images. Note that it permits the use of histochemistry as well as thin sectioning for transmission electron microscopy. See Figs. 10–5 and 10–6.

reduced amount of accelerator, the polymerization is complated in 3–6 days at 60°C. When infiltration is difficult, it often helps to cut the tissue into smaller pieces and to replace the epoxy resin every 12 hr with fresh (less viscous) resin.

After embedment and polymerization of the embedding medium, the tissue blocks are cut in the usual manner for light and electron microscopy. We have successfully used such histochemical stains as periodic acid Schiff, nile blue, sudan black B, toluidine blue, and Feulgen reagent on thick sections of scanned specimens for observation in the light microscope. Characteristic histochemical reactions are observed in every case, a feature which adds to the identification of specific scanned structures and provides localized histochemical information. Thin sections observed by TEM can be used to specifically identify structures which appear in detail in the scanning electron micrographs (Figs. 10–5 and 10–6). Frequently, during SEM observation of intracellular organization, it is advantageous to obtain TEM profiles for specific identification of structures. For example, objects which seem to be cell nuclei may be ultimately

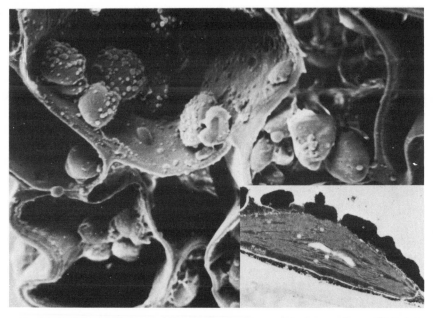

Fig. 10–5. Gold-palladium-coated mesophyll cells (*Sarracenia*) show numerous elliptical internal structures which are frequently covered with minute droplets. These structures are PAS positive and, when sectioned for the TEM, are identified as chloroplasts with associated secretory droplets projecting into the tonoplast vacuole (insert). Note the "negative" appearance of the unosmicated membranes. 5 kV. 2,850×; the insert, 12,500×.

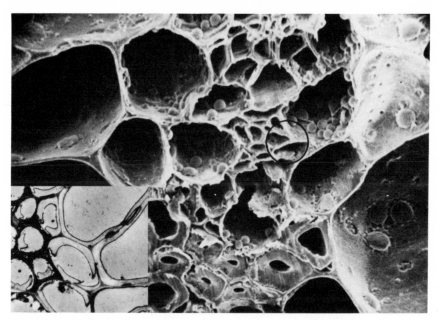

Fig. 10–6. A vascular bundle in the leaf of *Sarracenia*. Preparation of this metal-coated specimen for thin sectioning provides specific information on the vascular elements (TEM insert). 5 kV. 1,900×; the insert, 5,000×.

identified as secretory droplets when the same cell is observed in TEM profiles.

Another natural consequence of this kind of evaluation of the scanning image is a realization that the carbon-metal coat evaporated onto the specimen surface is not always faithful to the topography of the specimen. Larger structures (400 to 1000 Å) are revealed by the thin metal film (75 to 250 Å) on the surface of the SEM specimen; however, many important features of the cell interior are blanketed by this coating. Structures such as ribosomes, Golgi, pinocytotic vesicles, and membranous confluences can be obscured by a metal coating 200 Å thick.

Every scanning electron microscopist strives, sooner or later, to observe uncoated tissue (Pfefferkorn *et al.*, 1972). The direct observation of fresh specimens has been carried out by many investigators who have faced the same problems of desiccation and charging which limit resolution and the use of high magnifications (Heslop-Harrison, 1970; Falk *et al.*, 1970; Mozingo *et al.*, 1970; Echlin, 1971). For information on uncoated specimens, the reader is referred to Howden and Ling in this volume.

Electrolyte solutions and antistatic agents have been employed to supply biological specimens with additional conduction properties; however, these reagents do not fortify the tissue against the vacuum in the specimen chamber, nor do they improve contrast. We have found that the use of relatively long fixation for tissue stabilization, glycerol infiltration, combined with the treatment of the tissue with conductive material, can be used to produce specimens which are conductive and can be observed without metal coating. In addition, these specimens have satisfactory stability in the microscope, and beam penetration is reduced in thin areas. This has been described as the "tissue conductance technique" (TCT) (Panessa and Gennaro, 1973).

UNCOATED SPECIMENS

Tissues fixed in buffered glutaraldehyde (3 to 6%) for two days or more are rinsed several times in buffer to remove excess fixative. Following the buffer wash, the specimens are rinsed in two washes of distilled water and placed in an aqueous 2% potassium iodide-iodine solution (Johansen, 1940), which is prepared as given below.

potassium iodide	2.0 gm
iodine crystals	0.2 gm
distilled water to make 100 ml	

After 2 min the idoine solution is removed and the specimens are rinsed in distilled water until the supernatant is clear. A dilute lead acetate solution (1 part stock/5 parts distilled water) is filtered (0.45 μm pore size) and applied to the specimens for 3 min (Fig. 10–7).

lead acetate	1.2 gm
sodium hydroxide pellets	4.0 gm
distilled water to make 100 ml	

This stock solution is diluted prior to use.

After treatment with the lead solution, the specimens are rinsed for 3 min in distilled water. If the fixation process has hardened the tissue sufficiently, it may be placed in 50% glycerol for 30 min to several days and examined after glycerol substitution. This procedure has yielded excellent results in our laboratory (Figs. 10–8 and 10–9). If the tissue is unusually fragile or has not been allowed to harden adequately, it may be dehydrated in ethanol in the usual manner. This may be followed by substituting the ethanol with amyl acetate, which is miscible

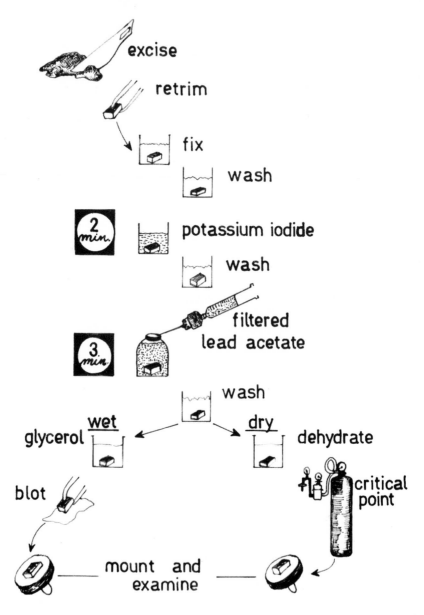

Fig. 10–7. Flow chart of the Tissue Conductance Technique (TCT). Note its compatability with general methods of tissue preparation for scanning electron microscopy (freeze-drying is not shown). See Figs. 10–8 and 10–9.

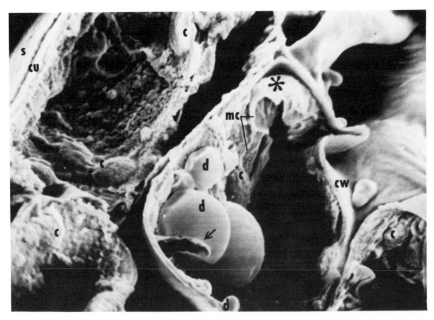

Fig. 10–8. Uncoated specimens prepared by the TCT show numerous droplets and cytoplasmic channels seen in profile in these epidermal cells of *Sarracenia*, which are also seen in transmission electron micrographs of these cells. The trimming process has cut through a secretory droplet (arrow). In places, the cytoplasm and plasmalemma have been pulled away from the cell wall (*). *d,* droplets; *c,* chloroplasts; *cw,* cell wall; *cu,* cuticle; *mc,* membranous channels; *s,* outer surface of the leaf. 10 kV. 4,000×.

with liquid CO_2. In this state, the tissue may be dried in liquid/gaseous CO_2 according to the critical point drying method (Anderson, 1951). This method maintains delicate structural relationships, although some shrinkage occurs during dehydration (Echlin, 1971; Boyde, 1972). The specimens prepared by the "conductance technique" and dried by the critical point drying method are fastened to the specimen stub with silver-conducting paint and examined without further treatment.

Specimens infiltrated with glycerol are examined wet. They are removed from the glycerol, the bottom surface blotted on a lint-free cloth, attached to a specimen stub with silver paint, and examined with the SEM. Higher accelerating voltages can be employed for examining these specimens. The authors have observed biological specimens prepared by the TCT with the scanning electron microscope under accelerating voltages from 5 to 20 keV and found specimens to be stable for up to 5 hr under the beam. The use of reduced raster for focusing is not recommended, because raster reduction concentrates the beam and increases

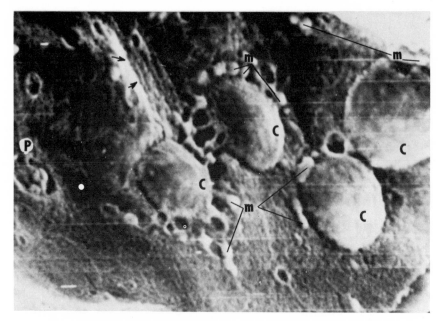

Fig. 10–9. Uncoated specimens yield information on mitochondrial (m)-chloroplast (c) associa-tion. Note the parallel arrays in the cytoplasm (arrows) similar to profiles of rough endoplasmic reticulum in TEM micrographs. P, primary pit field. 7 kV. 9,500×.

local heating. This increased local heating may cause shifting of the tissue beneath the metal coating, cracking of the metal surface film, or in some cases, what seems to be melting of the metal coating (Fig. 10–10).

The TCT was developed in the authors' laboratory when it became obvious that the surfaces of plant and animal tissues coated with even thin metal films (100 to 200 Å) seem to lose details compared to the surfaces of the tissue examined without coating. However, uncoated specimens may not be observed for long periods of time or at high mag-nifications. By using TCT, on the other hand, it has been possible to observe the cell interior and even visualize relatively small cellular

Fig. 10–10. This micrograph of the metal filmed (gold-coated) terminal bristles on a seta of the toe of the lizard (*Gekko gecko*) demonstrates the type of damage which may occur at high magnifications or reduced raster of minute, metal structures. These keratin bristles are quite small (1,000 Å in diameter) and each ends in a terminal cup (1,500 Å in diameter). At low magnifications, the bristles are erect and terminal cups facing upward. However, when the magnification was greatl,' reduced (raster size reduced) the bristles no longer appeared upright, but rapidly became "Wilted". Since this deformation was permanent, it is unlikely that this re-sulted from charging. 20 kV. 4,400×. Insert, 10,000×.

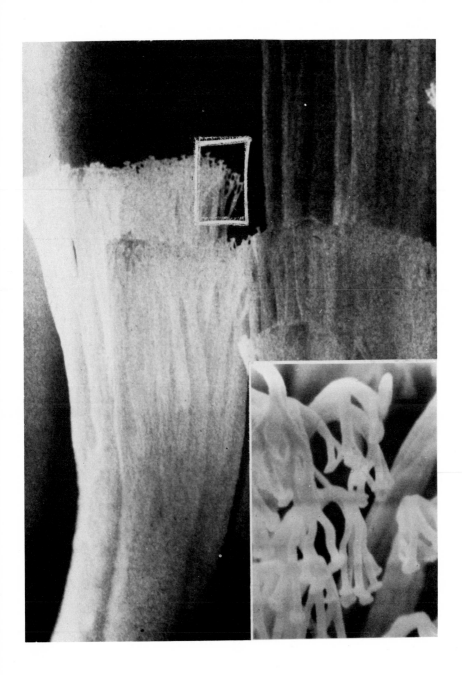

organelles and minute membranous channels (Fig. 10–8). Positive identification of all of these structures has been made from TEM images prepared from embedded scanning specimens. Botanical specimens examined after using TCT remain green and turgid after several hours of examination within the SEM column. When specimens prepared this way are embedded and sectioned for control observation with the TEM, no coating or precipitate can be found on the tissue. The fact that the tissue can be removed from the column, dissected, and immediately returned to the column without further treatment is another considerable advantage of the TCT (Panessa and Gennaro, 1973).

Of primary interest in the microscopy of the cell interior is the disposition of the membrane systems. Positional relationships such as those existing between mitochondria and areas of energy consumption are acknowledged to be significant. The idea of functional continuities within the intracytoplasmic membrane system has been approached (Crotty and Ledbetter, 1970; Grove et al., 1970; Morré et al., 1971). It is attractive to conceive of intracellular interactions being controlled in some manner by membranous confluences; however, although these are occasionally demonstrable with the TEM, only scanning electron microscopy will readily show an en face image that presents the topography in a total manner. SEM work of this kind may forecast great contributions to studies of the biology of the cell.

REFERENCES

Anderson, T. (1951). Techniques for the preservation of three dimensional structure in preparing specimens for the electron microscope. Trans. N. Y. Acad. Sci., Ser. II, 13, 130.

Beaman, D., and Isasi, J. (1972). Electron Beam Microanalysis, pp. 1–80. American Society of Testing Materials, Philadelphia.

Boyde, A. (1972). Biological specimen preparation for the scanning electron microscope. In: Scanning Electron Microscopy. Proc. 5th Ann. Scan. Electron Micros. Symp., p. 257. I.I.T. Research Institute, Chicago.

Crotty, W. J., and Ledbetter, M. C. (1970). Program (demonstration) 10th Ann. Meet. Amer. Soc. Cell Biol. San Diego.

Echlin, P. (1971). The examination of biological material at low temperatures. In: Scanning Electron Microscopy. Proc. 4th Ann. Scan. Electron Micros. Symp., p. 225. I.I.T. Research Institute, Chicago.

Echlin, P., and Hyde, P. (1972). The rationale and mode of application of thin films to non-conducting materials. In: Scanning Electron Microscopy. Proc. 5th Ann. Scan. Electron Micros. Symp., p. 137. I.I.T. Research Institute, Chicago.

Falk, R., Simmons, E., and Cutter, E. (1970). Scanning electron microscopy of developing plant organs. *Science* **168**, 1471.

Grove, S. N., Bracker, C. E., and Morré, D. J. (1970). An ultrastructural basis for hyphal tip growth in *Pythium ultimum. Am. J. Bot.* **57**, 245.

Guttman, H., and Styskal, R. (1971). Preparation of suspended cells for SEM examination on internal cellular structures. In: *Scanning Electron Microscopy. Proc. 4th Ann. Scan. Electron Micros. Symp.*, p. 265. I.I.T. Research Institute, Chicago.

Heslop-Harrison, Y. (1970). Scanning electron microscopy of fresh leaves of *Pinguicula. Science* **167**, 172.

Idle, D. (1971). Preparation of plant material for scanning electron microscopy. *J. Microscopy* **93**, 77.

Johansen, D. (1940). *Plant Microtechnique.* McGraw-Hill, New York.

Morré, D. J., Mollenhauer, H. H., and Bracker, C. E. (1971). Origin and continuity of Golgi apparatus. In: *Results and Problems in Cell Differentiation. II. Origin and Continuity of Cell Organelles* (Reinert, T., and Ursprung, H., eds.), pp. 82–126. Springer-Verlag, Berlin-Heidelberg-New York.

Mozingo, H., Klein, P., Zeevi, Y., and Lewis, E. (1970). Venus's flytrap observations by scanning electron microscopy. *Am. J. Bot.* **57**, 593.

Mumaw, V., and Munger, B. (1971). Uranyl acetate as a fixative. *Proc. 29th Ann. Meet. Electron Micros. Soc. Amer.* (Arceneaux, C. J., ed.), p. 490. Claitor's Publishing Division, Baton Rouge, Louisiana.

Padawar, J. (1971). Mast cells: A scanning electron microscopic study. *Anat. Rec.* **169**, 394.

Panessa, B. (1971). Determination of cellular organization from scanning electron micrographs. *Proc. Amer. Soc. Cell Biol.*, p. 215. New Orleans.

Panessa, B., and Gennaro, J., Jr., (1972). Preparation of fragile botanical tissues and examination of intracellular contents by SEM. In: *Scanning Electron Microscopy. Proc. 5th Ann. Scan. Electron Micros. Symp.*, p. 327. I.I.T. Research Institute, Chicago.

Panessa, B., and Gennaro, J., Jr. (1973). Use of potassium iodide/lead acetate for examining uncoated specimens. In: *Scanning Electron Microscopy. Proc. 6th Ann. Scan. Electron Micros. Symp.*, p. 395–403 I.I.T. Research Institute, Chicago.

Pfefferkorn, G. (1970). Specimen preparation techniques in scanning electron microscopy. In: *Scanning Electron Microscopy. Proc. 3rd Ann. Scas. Electron Micros. Symp.*, p. 89. I.I.T. Research Institute, Chicago.

Pfefferkorn, G., Gruter, H., and Pfautsch, M. (1972). Observations on the prevention of specimen charging. In: *Scanning Electron Microscopy. Proc. 5th Ann. Scan. Electron Micros. Symp.*, p. 147. I.I.T. Research Institute, Chicago.

Wodzicki, T., and Humphreys, W. (1973). Maturing pine tracheids: Organization of intravacuolar cytoplasm. *J. Cell Biol.* **56**, 135.

11. WOOD

Karl Borgin

Department of Wood Science, Stellenbosch University
Stellenbosch, South Africa

INTRODUCTION

In order to exploit the potentialities of the scanning electron microscope in wood science and to be able to use the most suitable methods for preparing wood for scanning electron microscopy, it is necessary to fully appreciate and understand the advantages and disadvantages of scanning electron microscopy compared with light microscopy and transmission electron microscopy.

Neither the transmission nor the scanning electron microscope can identify colors. This is the reason for the difficulties encountered in locating materials such as high polymers (Timmons, 1971) and adhesives (Collett, 1970) which have penetrated into the structure of the wood. Reduced to its most elementary form, a picture formed by a transmission electron microscope is roughly a transparent silhouette of the original section or particle. Structural details are seen only when they correspond to differences in electron-optical densities. Surface topography can in most cases be seen only after special oblique shadowing with metal vapors has been carried out on the original samples or their replicas.

In contrast, scanning electron microscopy clearly reveals the topography and *surface* morphology of any suitably prepared material, but will not show any structural detail of the interior of a sample (see Nei, in this volume, for an opposite view). Although the scanning electron microscope has proved its superiority in the study of surface structures, and demonstrates an amazing depth of field, which makes it possible to study materials in a sharp, three-dimensional view, it is limited mostly to the study of surfaces.

A scanning electron microscope can probe very deeply inside a material if it contains openings, fissures, pores, or cavities such as the cell lumen of wood fibers. The observations of such openings, however, cannot be carried out below the surface whether this surface is an external surface obtained by machining the wood or an internal surface such as the helical reinforcement of the inside of certain wood tracheids. Although structures such as tyloses (Ishida and Ohtani, 1969; Sachs *et al.*, 1970) or details of the wood anatomy (Borgin, 1972b) can be studied deep inside the wood with the scanning electron microscope, the wood tissue must be dissected in such a way that it becomes, for the purpose of investigation, an external surface.

Both light microscopy and transmission electron microscopy can probe the interior of a sample, since differences in structure are in most cases accompanied by differences in their light-optical or electron-optical properties. These differences make observations, measurements, and photography possible. Most samples used in light microscopy and transmission electron microscopy are almost transparent. If these samples had been opaque, such techniques would reveal the outline of the sample only as a shadow. The interior of a sample cannot be probed, and therefore studied, with a scanning electron microscope, because specimens for scanning electron microscopy are usually quite opaque. This opacity is further accentuated by the conventional use of metal coating or shadowing.

The coating of wood with gold or gold-palladium, which is used in the usual coating techniques, changes the wood surface from an organic material, with differences in electron-optical properties, to a uniform and homogeneous metal with respect to ability to absorb, reflect, or scatter electrons. The only differences between the various structures of metal-coated wood samples which can be seen in a scanning electron microscope are those in shape and form.

The study of wood with a scanning electron microscope comprises primarily observations, measurements, and photography of the micromorphology of only the *surface* of the sample. This, which in itself could be a disadvantage, can be turned into an elegant method for structural studies if wood and wood fibers can be separated into distinct structural components by mechanical, physical, or chemical methods.

The dissection of the internal structure of wood into its components must therefore be carried out prior to its observation with the scanning electron microscope, since no detail is visible through the metal-coated surface. The smallest structure which can be seen with a scanning electron microscope is identical to the smallest structural element into which the material can be disintegrated or the smallest structure of a three-dimensional form already present on the surface. This is the most important difference between scanning electron microscopy and all other microscopic methods for the study of the structure of a specimen.

The successful study of wood and wood fibers with the scanning electron microscope depends upon the suitable methods for the preparation of the sample. This statement seems to be contrary to the usual concept that scanning electron microscopy requires less preparation of the sample than conventional transmission electron microscopy and light microscopy. Both statements are, however, valid under different circumstances.

If the fractured surface of wood stressed to failure is studied with the scanning electron microscope, the preparation of the sample is extremely simple; the preparation of the same sample for studying with the transmission electron microscope is almost impossible because of its extremely irregular surface. However, if the fibrillar or laminar structure of the cell wall needs to be studied with the scanning electron microscope, the fibrils and the laminae must somehow be exposed so that they form at least part of the surface of the sample.

The success of the study of the internal structure of wood with the scanning electron microscope therefore depends upon how successful one can be in splitting, disintegrating, fracturing, or otherwise separating the material into its structural components which will make the various structural elements part of the newly created surface.

The borderline between different anatomical elements of wood tissue, the transitional areas between various layers of the cell wall, and the interfaces between microfibrils and embedding matrix are exposed to stress when wood is exposed to external forces. When the wood is stressed to failure, the structure may break up in these areas, and the various components of the structure and ultrastructure can be made visible in the scanning electron microscope. Cohesive failure in the highly stressed areas of a structure can be utilized as a method not only for studying the structure itself but also for providing valuable information on the mechanisms of failure.

METHODS

The preparation of wood samples for scanning electron microscopy is comprised of two major steps. The first step consists of preparing the sample so that maximum information can be obtained. The second step consists of shaping, trimming, and mounting the sample for insertion into the scanning electron microscope. The first step in the preparation of wood samples for scanning electron microscopy depends upon the type of information to be obtained from the investigation. According to this definition, most wood samples can be divided into three categories:

(1) Wood in its natural, unmodified state.

(2) Wood which has been deliberately modified or changed as the result of exposure to laboratory experiments carried out for a specific purpose.

(3) Wood which has been modified or changed as a result of interaction with the environment.

Natural, Unmodified Wood

Wood is examined in the scanning electron microscope mainly in its natural, unmodified form to study either its anatomy or the structure of the cell wall (Brown and Baker, 1970; Collett, 1970; Echlin, 1968; Findlay and Levy, 1969, Meylan and Butterfield, 1972; Troughton and Donaldson, 1972). For this purpose, the wood is usually cut into cubes varying in size from a few millimeters to the maximum size which can be accommodated in the sample holder of the scanning electron microscope.

From a suitable size block or plank, cubes of ~10 mm are cut out with a chisel (Fig. 11–1c) or cut into strips of the same transverse section which can be placed in the microtome holder. A much faster method is to cut out cylindrical plugs of ~10 mm in diameter (Fig. 11–2), which are placed directly in the microtome either dry or after softening by boiling in water for 1 hr.

Although a microtome is the most suitable cutting instrument, surprisingly good results have been obtained with commercial razor blades

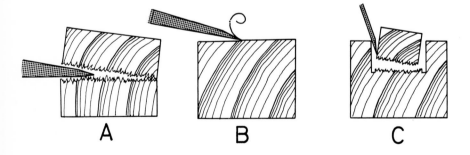

Fig. 11–1. Typical methods for preparing various surfaces for study with the SEM: (A) Cleaved sample to obtain a fractured surface; (B) microtomed sample to obtain a smooth, flat surface; (C) sample removed with a chisel to obtain an untreated surface in its natural state.

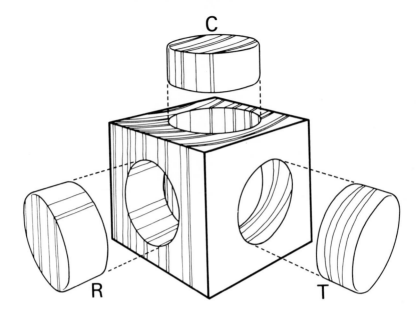

Fig. 11–2. Samples are often prepared using a plug cutter to expose three different types of surfaces: radial (*R*), tangential (*T*), and cross section (*C*).

(Brown and Baker, 1970; Findlay and Levy, 1969). Normally only one side of the cube is studied, and only this side is prepared by slicing it with a microtome (Fig. 11–1b). The other sides of the cube are cut with a scalpel and trimmed to size. Striking micrographs can be obtained by observing the cubes in perspective. Under such conditions, two or three sides must be perfectly cut. The investigations of Meylan and Butterfield (1972) are examples of successful techniques of this kind.

The condition of the microtome knife is important. Knife marks appear on the cut surface, although these do not ruin the informative value of the sample, they are detracting. Typical knife marks are seen on the otherwise satisfactory electron micrographs of *Pinus radiata* (Fig. 11–3); such marks are visible on most published micrographs of this kind.

The wood can be cut dry or wet after being boiled to soften the tissues. The technique of boiling small wood samples prior to microtomy is most successful in preparing smooth surfaces for scanning electron microscopy. The surface of the wood which has been cut while wet does not remain flat, although it is still smooth after drying. Because of the differences in the shrinkage of parts having different densities of the same sample (e.g., across the annual growth rings after the wood

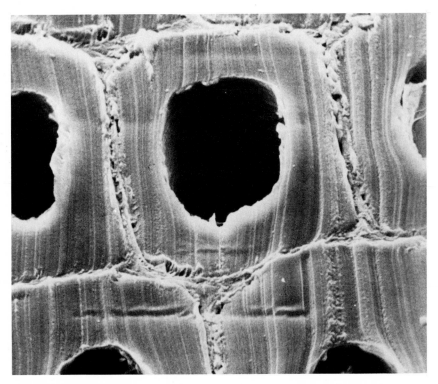

Fig. 11–3. Cross section of *Pinus radiata* cut from a frozen specimen with a microtome. Note knife marks in the direction in which the sample was cut. Deformation due to drying or exposure to high vacuum is negligible with hard materials such as wood. 6,200×.

has been dried), a slightly corrugated surface can sometimes be observed. This can be avoided by cutting the wood while dry, but the requirements for a sharp knife will be even more stringent in this case. A glass knife, as used in ultramicrotomy, gives excellent results, but can cut only rather small samples.

The author has found that the best method for cutting wood with a microtome is to use a sample which has never been dried or dehydrated. Such samples can be obtained only from recently felled trees; these samples are kept in water at 0 to 5°C.

There are differences in the viscoelastic properties of undried and drastically dried woods. Most commercial timbers have been drastically dried. The usual problems of bending instead of cutting the thin-walled fibers of dried wood are rarely encountered in natural, hydrated wood.

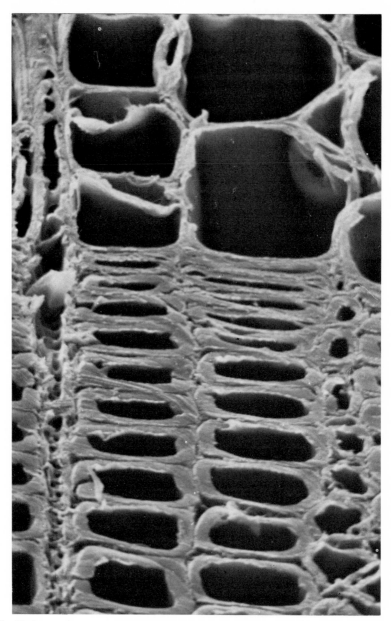

Fig. 11–4.

(a) Cross section of Norwegian pine (*Pinus sylvestris*) at the borderline between two year rings, 1,344×.

It is useful to describe the different surfaces studied by microscopic techniques. Fig. 11–2 shows how samples removed from a stem or a plank can have distinctly different surfaces. A transverse section (Fig. 11–4) can be cut only in one manner, but the longitudinal sections are either tangential (Fig. 11–5) or radial (Fig. 11–6), with quite different anatomical characteristics. The cutting or shaping of wood by knives or microtomes is more suitable for the study of the anatomy of the wood than for the study of ultrastructure. By slicing the wood in successive layers with a knife or a microtome, the anatomy has been studied in three dimensions by Meylan and Butterfield (1972) and others.

The ultrastructure of wood consists mainly of the laminar and fibrillar structure of the cell walls of fibers, tracheids, and vessels. Little information can be gained from microtomed surfaces or the exposed inside of the different cells, since the ultrastructure is concealed behind the surface. However, the micromorphology is revealed in striking clarity.

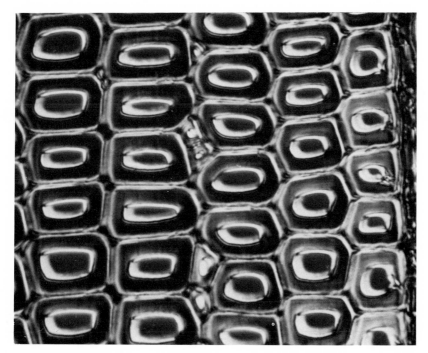

(b) Cross section of pine (*Tsuga canadensis*) observed with Nomarski interference technique. The microtomed section was untreated and unstained .

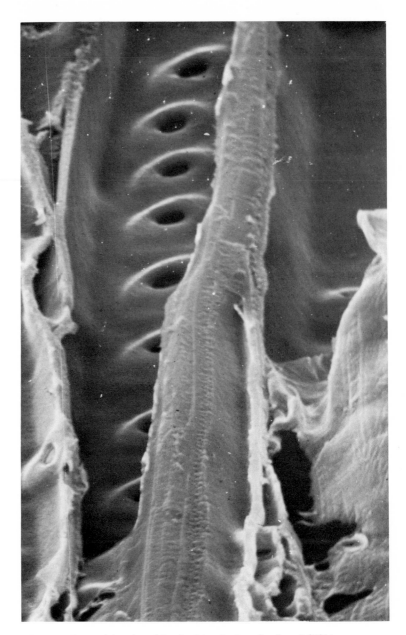

Fig. 11–5. Tangential section of *P. sylvestris*, microtomed surface. 1,380✕.

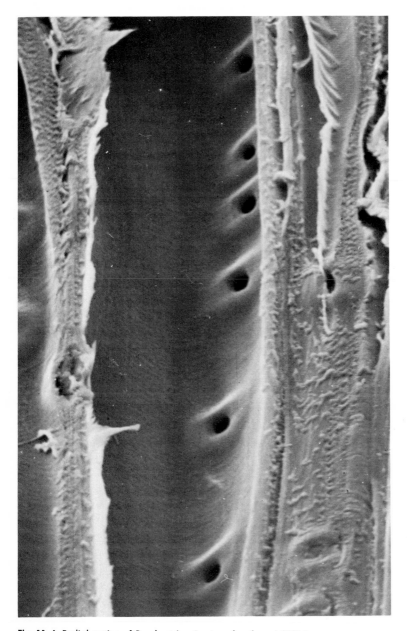

Fig. 11–6. Radial section of *P. sylvestris,* microtomed surface. 1,392×.

Preparation for Maximum Information

Wood can be modified either to study the effect of the modification of the structure itself or to study the mechanism of the process causing the modification. Scurfield and Silva (1969) have underlined the value of the scanning electron microscope in the examination and interpretation of the effects of applied physical and chemical treatments on the surface architecture of plant structures. Wood fractures have been used to study both the mechanism of fracture (Borgin, 1971c; Borgin and Van Zyl, 1971; Keith, 1971) and the behavior of the ultrastructure under stress (Borgin and Corbett, 1972; Kennedy, 1970).

In order to study the architecture of the cell wall, its ultrastructure must somehow be revealed by opening up or dissecting the internal structure in such a way that the interior is fully exposed. The coherent cell wall must be broken up into its structural components by creating external surfaces from the interior by various types of fractures. When wood fails under stress, different types of fractures can be observed. When wood is stressed to failure under controlled laboratory conditions, it can exhibit pure tensile, shear, and compression failures (Fig. 11-7). Any composite material will reveal its components in the scanning electron microscope after undergoing such failures. The different types of failures usually take place in the highly stressed borderlines between different components.

Wood, being a natural, complex composite, fails exactly in the manner stated above. When wood fails under tension, shear, or compression, the cell wall layers and the microfibrils are seen in three dimensions in the scanning electron microscope (Borgin, 1971c; Borgin and Corbett, 1972; Borgin and Van Zyl, 1971; Keith, 1971). A typical example of this is shown in Fig. 11-8, where the laminar cell wall layers as well as the microfibrils are clearly visible. This cannot be accomplished by using any other microscopic method. The mechanism of failure and the ultrastructure made visible depend not only upon the type of stresses but also on the rate of applying the stress as well as on the temperature and moisture content.

By controlling the conditions stated above, more information can be obtained and significant conclusions can be drawn. The information gained from natural fractures, where the exact type of stress is not known, is rather limited.

The study of wood which has been modified experimentally by mechanical or chemical means offers probably the most versatile and profitable field of research with the scanning electron microscope. The purpose of such investigations is either to study the effect of the processes

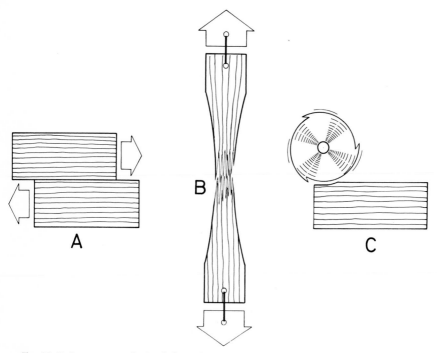

Fig. 11–7. Fracture or cohesive failure of wood to produce surfaces for study with the SEM should be accomplished under controlled conditions: (A) shear parallel or perpendicular to the grain; (B) tension parallel to the grain; (C) sliced by machining.

themselves or to test different processes for exposing structural features which can be studied with the scanning electron microscope.

The functions and efficiency of machining operations such as cutting, sawing, planing, and drilling are of great industrial importance, and can be studied in detail with the scanning electron microscope much better than by any other method. Chemical pulping of wood and bleaching and boiling of wood pulp for paper making have a specific and significant effect on the structure of the single wood fibers. The delignification of the wood tissue and the fibrillation of paper making fibers (Gonin and Taylor, 1972; Koran, 1970) can be studied directly with the scanning electron microscope without any special preparations. Suspensions of wood or cellulose fibers can be placed directly on the specimen stub and can, after drying, be coated without any further preparation. Generally the fiber suspension contains enough hemicellulose, pectins, and decomposed cellulose to act as a thin layer of adhesive to which the fibers or fiber fragments will adhere.

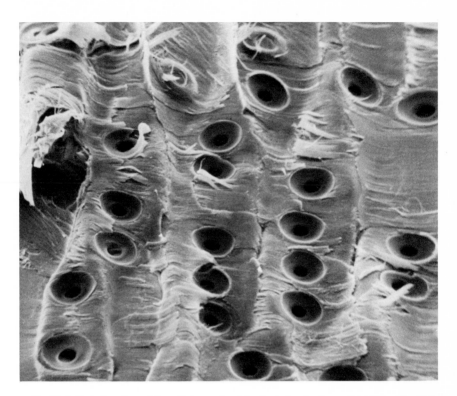

Fig. 11–8. Typical surface of *P. radiata* produced by fracturing in shear along the grain. Note failure of middle lamellae, separation of various layers of the cell wall, splitting of the laminar structures into fibrils and microfibrils, and distinct micromorphology of the dome-shaped pits. The detailed information obtained from such fractured surfaces is difficult to obtain from wood cut with a microtome. 2,400×.

As has been mentioned previously, in order to study the structure and ultrastructure of wood and wood fibers, the interior of the wood tissue and the wood fibers must be exposed by fracturing the structure. A number of processes are available to accomplish this. Small samples of wood can be stressed to failure, preferably under controlled conditions in a testing machine where stress and the rate of stress application as well as the temperature can be controlled and varied to suit the experiment (Fig. 11–7).

External stresses are normally applied in the three main directions of the wood structure—that is longitudinal, radial, and tangential. The mechanism of failure in each of these three directions is different, and the wood fractures differently in each, depending upon the rate of stress. The middle lamella, for instance, responds differently to a given stress than does the fibrillar structure of the cell wall (Borgin, 1970 and 1971a).

The study of the failure of materials such as metals and high polymers will supply information on the structure of the material itself. The information on the structure and ultrastructure of wood obtained by using this method is unobtainable by any other means. This method reveals not only the details of the structure but also how this structure behaves under stress. The scientific significance of this type of information is considerable, since wood probably provides the strongest structure possible with the minimum amount of material. It is therefore of considerable importance to know how this natural composite material behaves when compared with synthetic materials, which are making spectacular progress in modern technology.

The sampling of the fractured surfaces can be carried out according to the technique shown in Fig. 11–9. Since the surface must neither be touched nor interfered with, small plugs are drilled out or small cubes are chiseled from the wood. Often the wood is fractured into small fragments or splinters, so that it can be studied directly with the scanning electron microscope after necessary coating.

Naturally Modified Samples

Wood exposed to the environment undergoes a series of changes. The most noticeable is the slow erosion of the surface of the wood (Borgin,

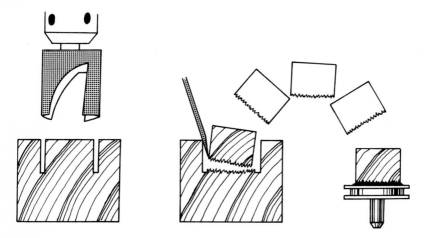

Fig. 11–9. Preparation of circular samples using a motor-driven, hollow plug cutter in order to obtain specimens from naturally or experimentally made surfaces which are studied in their original, undisturbed condition—for example, weathered or fractured samples. After cutting, the samples are pried loose and mounted directly on the specimen stub as shown here.

1971a and b; 1972a and c). While serial slicing of the wood tissue with a microtome will give a smoothly cut surface, controlled or natural erosion will often expose the various components of the ultrastructure as the erosion proceeds, layer by layer, deep into the wood. As in the case of cohesive failure and fracture, the forces causing the erosion will attack the weakest components first, which in most cases will expose the more resistant parts such as microfibrils.

The rate of the erosion is important. When wood is exposed to sandblasting with abrasive particles, the composite structure behaves as a homogeneous material, and the wood is eroded without revealing any structural elements. Very slow erosion produces quite different results, and natural weathering is mainly such a process (Borgin, 1970, 1971b, 1972a; Kühne et al., 1970; Sell and Leukens, 1971). Borgin (1971a) has shown that the structure of wood breaks down during weathering processes almost in the reverse order from which it is built up. Middle lamellae, cell wall layers, and microfibrils slowly separate from each other and disintegrate (Fig. 11–10). This separation facilitates the study of structural elements and the relative strength, durability, and resistance

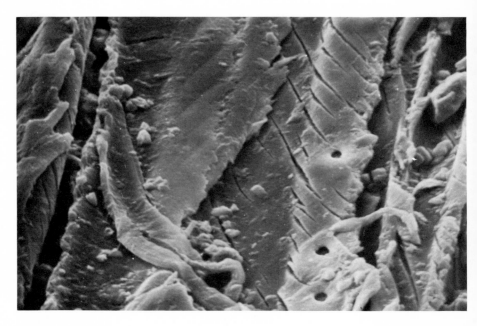

Fig. 11–10. Effect of weathering on *P. radiata* under South African climatic conditions. Eroded and fractured surface with fissures in the tracheids following the direction of the microfibrils in the cell wall is apparent. Natural surface. 1320×.

of wood to deterioration with the scanning electron microscope. Weathering is therefore another method of dissecting wood into its components, thereby providing valuable information.

The wood which has been modified by interactions with the environment can be studied for changes which have taken place on the surface or in the interior of the wood structure. Typical *surface* phenomena are changes of the exterior of the wood caused by weathering and by living organisms (Fig. 11–11). Most changes in the *interior* of the wood are also caused by living organisms; weathering is largely a surface phenomenon (Borgin, 1971a and b, 1972a).

The study of the weathering of the surface of wood is a typical example of the use of scanning electron microscopy where no other technique can supply the same information. One reason is that the surface as studied with the scanning electron microscope is not interfered with prior to the insertion of the sample into the specimen holder of the microscope. The surface of the sample must be completely intact, and this is accomplished by drilling plugs with a plug cutter, as shown in Fig. 11–9. Chiseling is less satisfactory, since very old samples of reduced mechanical strength may be compressed during this operation.

Old surfaces of weathered wood can provide information on the structural changes and the remarkable stability of wood under certain conditions (Borgin, 1971b, 1972a; Kühne et al., 1970). Wood samples up to several million years old have been studied with the scanning electron microscope (Wayman et al., 1971), and information on the durability of the structure of wood has been obtained (Fig. 11–12). The study of fungi and other microorganisms on the surface of the wood reveals how the wood is decomposed by these agents. Insects, bacteria, and fungi can penetrate deep into the wood, and the results of this type of deterioration can be studied only by cutting samples with a microtome, as explained earlier. The procedure used to accomplish this is not easy, for two reasons. First, the structure of the wood sample has to be opened with a knife. Secondly, the exact region which has been destroyed by the microorganisms must be located. This often requires the cutting of a number of consecutive layers to find one which will reveal the presence of microorganisms or the effect of the enzymatic breakdown of the wood structure.

By using this approach, one can gain information on the structural damage caused by microorganisms and on the mechanism for the propagation of the fungal hyphae within the ultrastructure of the cell wall (Findlay and Levy, 1969). In the same way that mechanical stresses will fracture the cell wall in its weathered part, the enzymes from fungi and other microorganisms will find the path of least resistance and first de-

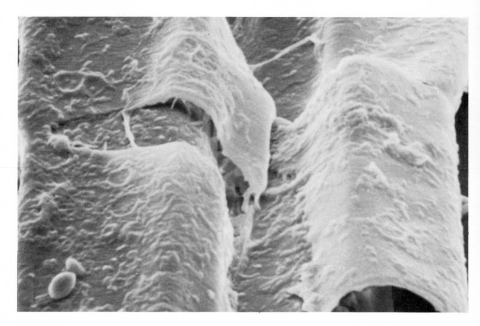

Fig. 11–11. Enzymatic breakdown of surface fibers of *P. radiata* caused by bacteria. The chemically and mechanically weaker components of the wood-lignin complex are attacked first, which results in the delamination of the cell wall into sheetlike laminae. Natural surface of sawed planks. 2496×.

stroy the part of the ultrastructure with the largest accessibility and/or the least chemical stability.

The technique used to study the effect of weathering of wood can also be used for investigating the stability and durability of surface coatings which must not be disturbed or interfered with before they are studied with the scanning electron microscope.

The penetration of adhesives and high polymers into the wood structure can be studied by cutting blocks containing the glue line and by slicing off the surface in a number of consecutive layers with the microtome. Since the scanning electron microscope does not distinguish colors, difficulties may be encountered in establishing the exact position and pathway for penetration. Also, the glue and other impregnants lack any recognizable structure. A case in point is the penetration of phenol adhesives or creosote preservatives into the wood. After penetration, they can be easily studied with the light microscope because of the strong color contrast. The use of the scanning electron microscope for the same purpose does not produce such striking results.

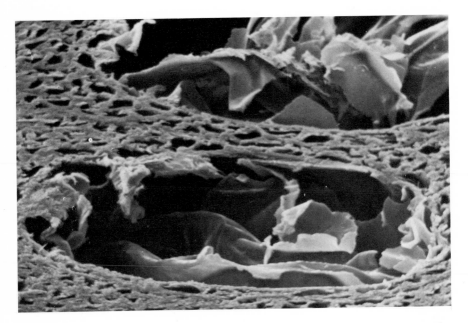

Fig. 11–12. The anatomical structure of the oak keel (*Quercus robur*) of a Viking ship buried for ~900 years in Oseberg (Norway) and now preserved and reconstructed in Oslo. The typical large vessels in oak containing tyloses are clearly visible, although some cell cavities are filled with an encrusting material because of the long contact with wet soil. The cell structure is slightly compressed. Microtomed surface. 576×.

Mounting and Coating of Samples

Because of the excellent mechanical properties of wood, the high vacuum encountered in the coating equipment and the scanning electron microscope has very little effect on the structure and morphology of the samples. Wood is a robust specimen not easily damaged, and no serious problems are encountered in drying in vacuum (Echlin, 1968). No pretreatment of the samples is therefore necessary, although a different solvent exchange system can be used when very delicate parts of the anatomical structure need to be studied. Collett (1970) used a solvent exchange system to preserve the fine structure of the torus of bordered pits of pines; Gonin and Taylor (1972) used the same method for preserving the structure of defibrillated wood fibers.

Generally the samples can be placed directly on the specimen holder without any preparation. A number of specimen holders or stubs are in use. Typical mounting stubs are shown in Fig. 11–13 (JEOL, Hitachi, Stereoscan). The mounting of the samples on the specimen stubs is

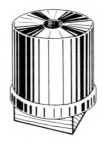

Fig. 11–13. Various specimen holders from different makes of SEMs showing samples of wood glued in the correct position. The majority of the holders are placed in the microscope column with the sample facing upward (right and left), but some holders are placed in such a manner that the sample faces downward (center).

simple. A small amount of synthetic adhesive (e.g., polystyrene or conductive silver paint) is placed on the stub and on one side of the wood sample. A light pressure is applied to fasten the sample firmly in position.

Very often the specimens are studied in a central electron microscope laboratory. Under such conditions it is important to be able to transport the samples without touching their surface. The plastic container shown in Fig. 11–14, which holds eight specimen stubs with the mounted samples, is most useful for this purpose. In the scanning electron microscope, all nonconducting specimens build up an electric charge as the electrons strike them. To dislodge the charge, the wood samples must be made conductive by depositing a layer of carbon or metal on their surface. Echlin (1968) has given details of the coating of different plant materials. However, wood requires a standardized treatment which can be repeated for general use.

Wood is extremely porous, and contains, in addition to visible and microscopic pores, a large amount of submicroscopic cavities. Gases and water vapors are held in the porous part of wood, and condensed water is held by strong forces in the capillaries. Both the degassing of wood samples prior to coating under vacuum and the pumping of vacuum in the specimen chamber of the scanning electron microscope can be prolonged processes. This is especially true for vacuum coaters which accommodate a large number of samples.

Because of the extreme porosity of the structure of wood, the coating

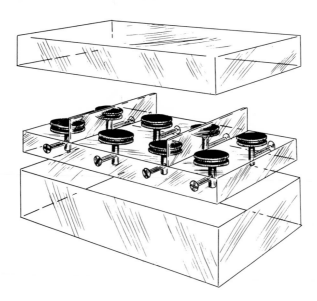

Fig. 11–14. Special containers for several specimen stubs. These prevent the surface of the wood from being disturbed or contaminated during storage or transport.

of the interior of the wood fibers and other cavities is difficult. The specimen holders inside the vacuum coater should be of the rotating type, and the tilt should vary during the coating process. Although coating with only gold-palladium yields excellent results, it is advisable to first deposit a thin layer of carbon.

The thickness of the conductive coating has been the subject of many discussions. Every investigator seems to have his own method, which he feels is the best (Brown and Baker, 1970; Echlin, 1968; Pfefferkorn, 1970). Generally, however, smooth surfaces such as those obtained using microtomy require a thinner coating than do fractured surfaces with needlelike fiber fragments, which are easily charged in the microscope. Very thin coatings can be used for microtomed specimens—so thin in fact that the color of the wood is still discernible. Fractured surfaces are coated to yield a true metallic surface.

Coatings varying in thickness from 50 to 500 Å have often been quoted in the literature; but since it is almost impossible to establish the thickness of the carbon or gold-palladium lays on porous wood, figures like these have little meaning. The best results are obtained by trial and error. If the wood is fastened to the specimen stub with a nonconducting high polymer, a strip of silver paint must be applied between the sample holder and the coated specimen.

OBSERVATION OF SAMPLES

Wood samples do not show significant damage during observation even after prolonged exposure to the electron beam. Excessive electron current may, however, cause heating and deformation of at least part of the sample. If possible, the microscope should be operated at 10 kV; however, 20 kV usually does no harm. The only serious problems are those encountered with fibers protruding from fractured surfaces. Because of the very low thermal conductivity and the tendency for electrons to accumulate on sharp points, these fibers are rapidly damaged by heat. This can be an annoying phenomenon, particularly when the fractured surfaces of single fibers are studied.

IMPORTANCE OF STUDYING WOOD

Wood is probably the material most widely used by man. Even today, in the age of new and important materials, wood is one of our most versatile and valuable structural materials. Since the science of composites has been applied most successfully in the fields of wood technology and timber engineering, the use of wood as a structural material in the form of glued laminated timber is more important than ever before, in spite of the spectacular developments in synthetic structural materials.

With the exception of the chemical conversion of wood or certain components of wood into plastics, rayons, films, adhesives, and surface coatings, the use of wood depends almost exclusively upon the microscopic and submicroscopic structure of the wood and the wood fibers. Wood is nature's own composite of an incredibly ingenious design. Its bulk structure—the arrangement of the wood cells and fibers into various types of wood tissues—and the architecture of the ultrastructure of each single wood fiber are masterpieces of structural engineering.

The excellent strength properties of wood are utilized not only when wood is used in its bulk form and as glued laminated timber structures but also when it is disintegrated into particles and fibers and reconstituted into particle board, fiberboard, and paper. Without the unique structural properties of each wood fiber, the worldwide pulp and paper industries could not exist.

Although the long, linear cellulose molecule is the ultimate source of the excellent strength properties of wood and wood fibers, it is the arrangement of the cellulose molecules into a microscopic and submicroscopic structure which is responsible for most mechanical, physical, and even some chemical properties of wood. The study of the structure of wood is therefore the most important single field of research in wood science and technology. The value of new and better techniques and methods for the study of the structure of wood is obvious.

Investigations on the structure and ultrastructure of wood and wood fibers had, prior to the development of the scanning electron microscope, been carried out by using various light microscopy techniques and transmission electron microscopy. The scanning electron microscope makes possible the study of micromorphology of wood, the structure and ultrastructure of the cell wall, and the single fibers.

I am thankful to Mr. Jon O. Seaman for making drawings and to Miss K. Corbett for preparing electron micrographs.

REFERENCES

Borgin, K. (1970). The use of the scanning electron microscope for the study of weathered wood. *J. Microscopy* **92**, 47.

Borgin, K. (1971a). The mechanism of the breakdown of the structure of wood due to environmental factors. *J. Inst. Wood Sci.* **5**, 26.

Borgin, K. (1971b). Why wood is durable. *New Scient. Sci. J.* **50**, 200.

Borgin, K. (1971c). The cohesive failure of wood studied with the scanning electron microscope. *J. Microscopy* **94**, 1.

Borgin, K. (1972a). Bestandighet og varighet av norsk trevirke. *Teknisk Ukeblad* **119**, 15.

Borgin, K. (1972b). Feinstruktur von Holzgewebe. *Umschau* **72**, 326.

Borgin, K. (1972c). The stability and durability of wood from pines grown in cold and warm climates. *Proc. 7th World For. Congr.* Buenos Aires.

Borgin, K., and Corbett, K. (1972). The behaviour of the ultrastructure of wood under extreme stresses. *Proc. South African Electr. Micr. Soc. Symp.*, p. 33–34.

Borgin, K., and Van Zyl, J. D. (1971). Structural changes in wood due to compression and densification. *Proc. South African Electr. Micr. Soc. Symp.*, p. 45–48.

Brown, F. L., and Baker, H. M. (1970). Scanning electron microscopy of mature Douglas-Fir early wood intertracheid pitting. *Wood and Fiber* **2**, 52.

Collett, B. M. (1970). Scanning electron microscopy: A review and report of research in wood science. *Wood and Fiber* **2**, 113.

Echlin, P. (1968). The use of the scanning reflection electron microprobe in the study of plant and microbial material. *J. Roy. Micros. Soc.* **88**, 407.

Findlay, G. W. D., and Levy, J. F. (1969). Scanning electron microscopy as an aid to the study of wood anatomy and decay. *J. Inst. Wood Sci.* **4**, 57.

Gonin, C. R., and Taylor, F. W. (1972). A scanning electron microscopic study of fibrillation of pulp fibers. *Proc. South African Electr. Micr. Soc. Symp.*, p. 37–38.

Ishida, S., and Ohtani, J. (1969). Study of tyloses using the scanning electron microscope. *Proc. 2nd Ann. SEM Symp.*, Chicago, p. 197.

Keith, C. T. (1971). The anatomy of compression failure in relation to creep-inducing stresses. *Wood Sci.* **4,** 71.

Kennedy, R. W. (1970). An outlook for basic wood anatomy research. *Wood and Fiber* **2,** 182.

Koran, Z. (1970). Surface structure of thermo-mechanical pulp fibers studied by electron microscopy. *Wood and Fiber* **2,** 247.

Kühne, H., Leukens, U., Sell, J., and Wälchli, O. (1970). Investigations on weathered wood surfaces. *Holz als Roh-u Werkstoff.* **28,** 223.

Meylan, B. A., and Butterfield, B. G. (1972). *Three-Dimensional Structure of Wood.* Chapman and Hall, London.

Pfefferkorn, G. E. (1970). Specimen preparation techniques. *Proc. 3rd Ann. SEM Symp.*, Chicago, p. 89.

Sachs, I., Kuntz, J., Ward, J., Nair, G., and Schultz, N. (1970). Tylose structure. *Wood and Fiber* **2,** 259.

Scurfield, G., and Silva, S. (1969). Scanning electron microscopy applied to a study of the structure and properties of wood. *Proc. 2nd Ann. SEM Symp.*, Chicago, p. 185.

Sell, J., and Leukens, U. (1971). Verwitterungserscheinungen an ungeschützen Hölzern. *Holz als Roh. u-Werkstoff.* **29,** 23.

Timmons, T. K. (1971). Polymer location in the wood-polymer composite. *Wood Sci.* **4,** 13.

Troughton, J., and Donaldson, L. A. (1972). *Probing Plant Structure.* McGraw-Hill, New York.

Wayman, M., Azhar, M. R., and Koran, Z. (1971). Morphology and chemistry of two ancient woods. *Wood and Fiber* **3,** 153.

AUTHOR INDEX

SUBJECT INDEX